Modern Physics
&Technology
for Undergraduates

Modern Physics &Technology
for Undergraduates

Lorcan M. Folan & Vladimir I. Tsifrinovich
Polytechnic University, New York, USA

Gennady P. Berman
Los Alamos National Laboratory, USA

World Scientific
New Jersey • London • Singapore • Hong Kong

Published by

World Scientific Publishing Co. Pte. Ltd.

5 Toh Tuck Link, Singapore 596224

USA office: Suite 202, 1060 Main Street, River Edge, NJ 07661

UK office: 57 Shelton Street, Covent Garden, London WC2H 9HE

British Library Cataloguing-in-Publication Data
A catalogue record for this book is available from the British Library.

530
F 663

MODERN PHYSICS AND TECHNOLOGY FOR UNDERGRADUATES

ISBN 981-02-4882-2
ISBN 981-02-4883-0 (pbk)

Printed in Singapore by Mainland Press

PREFACE

The main purpose of this book is to connect the standard introductory undergraduate course in physics with advances in modern science and technology. The first part of the book (Chapters 1–4) gives a very brief review of selected "hot" topics in modern physics and technology. The second part of the book (Chapters 5–7) includes problems at the introductory level of undergraduate physics, which are related to various hot topics of contemporary science. We emphasize that students can solve these problems without any knowledge of modern physics. They do not have to read the first part of the book. However, if a particular problem stimulates a student's curiosity, he or she may open the first part of the book to get a brief general tutorial on the subject. We hope that our book will help to revive interest in undergraduate physics and will help students to adjust their knowledge to advancing technology.

A huge chasm has developed between modern science and undergraduate education. The result of this chasm is that students who are graduating from college are unable to exploit the many opportunities offered by modern science and technology. Consequently, student interest in undergraduate physics is very low.

We propose to try to bridge this chasm between modern science and technology and undergraduate education. Modern science and technology widely uses the methods of classical physics, but these modern applications are not reflected in the problems on physics often suggested to students. Solving practical problems is a very effective way to inform students about contemporary science, to show the important relationships between modern and classical physics, and to prepare them for future activity in the modern technological environment.

We have prepared a set of problems based on some of the latest development in science and technology which can be solved using the classical physics accessible in a standard undergraduate program. We did this work in the Introductory Design and Science (IDS) Depart-

ment of the Polytechnic University (New York) in cooperation with
Los Alamos National Laboratory (LANL). Polytechnic University is
one of the oldest technical Universities in the country, widely recog-
nized as an excellent center for undergraduate education. The IDS
Department at Polytechnic University was created to develop new
approaches in scientific education. LANL is widely known as one of
the outstanding research centers in the world and it has an active
program to promote undergraduate science education.

L. M. Folan, V. I. Tsifrinovich and G. P. Berman

February 2003

CONTENTS

Chapter 1

INTRODUCTION

A huge gap exists between modern science and technology and standard undergraduate physics education. Typical problems we suggest to undergraduate students in a course of physics seem to have very little in common with contemporary discoveries and advances. At the same time, probably many scientists would agree that the level of introductory undergraduate physics is high enough to estimate and sometimes even compute some important features at the very edge of modern physics.

The main idea of our book is to promote reconciliation between modern science and undergraduate physics. We believe such a reconciliation will spark interest in both physics education and the current frontiers of science. The best outcome would be if it would help students to understand and appreciate modern scientific ideas in their future education and working lives.

The book is organized as follows: In the next three chapters, we present a short popular review of selected hot topics in modern science and technology. We hope both college students and college professors will find something interesting in our review. Chapter 2 presents the current picture of the fundamental elementary particles and fundamental interactions. We briefly describe the three generations of fundamental elementary particles, the compound elementary particles, and the gauge fields. Then we describe the transformations of elementary particles and some of the powerful accelerators designed to study the interactions between the particles. Next, we consider atomic nuclei and nuclear reactions, quantum properties of an atom including such delicate questions as Schrödinger cat states. For every topic we try to give basic information about the topic as well as current information from the frontiers of contemporary research. The selected topics certainly show some bias and are not intended to

1

be a complete popular review. The same principle, to an even greater extent is used in the next two chapters.

In Chapter 3 we present a few applications of modern physics. We briefly describe the principles of electron spin resonance and nuclear magnetic resonance and the ideas of spin refocusing (spin echo). Then we shift to contemporary ideas of single-spin detection including magnetic resonance force microscopy and scanning tunneling microscopy. We describe nanotubes as one of the frontier directions in nanotechnology. Then we discuss superconductivity including Josephson effects and the SQUID. From these "relatively old" phenomena we shift to frontier problems: superposition states in a nanometer scale "superconducting box," the Schrödinger cat state for a superconducting current, and superconducting magnets. We also mention application of physics to fighting natural disasters. As an example, we consider tsunami and contemporary ideas of tsunami warning. Finally, we explain the intriguing contemporary ideas of quantum computation and quantum teleportation.

In Chapter 4 we consider astrophysical phenomena. We describe the fusion reactions in a normal star, like our Sun. Then we consider a normal star's evolution. We describe the origin and properties of White Dwarfs, Neutron Stars, and Black Holes. Finally, we briefly consider some of the current space flight programs, their achievements and expectations.

Chapters 5 and 6 contain short reviews of the main topics from a standard introductory undergraduate course of physics and the problems. The suggested problems are associated with frontier scientific research. Some of the problems are directly connected to the topics discussed in the previous three chapters. To solve these problems a reader does not have to go through the earlier chapters. However, if a topic connected to a problem has excited you, we encourage you to read the corresponding section.

Chapter 5 contains a short review and problems based on an introductory undergraduate, calculus-based course on mechanics including traveling and standing waves and also sound waves, which are often considered in a separate course.

Chapter 6 contains a review and problems on electricity and magnetism, including electromagnetic waves and even optical phenomena, which often are also treated in a separate course on optics.

Chapter 7 contains hints to solution of the problems except for a few very simple problems.

Finally, the Appendices contain some useful data and formulas which can be used in the solution of the problems.

Chapter 2

ELEMENTARY PARTICLES, NUCLEI, ATOMS

2.1. Fundamental Elementary Particles

The contemporary scientific picture of the Universe is rather amazing. The Universe is filled with fields. There are three "generations" of fields responsible for the fundamental elementary particles. The "first generation" is the most important. It consists of the "electron neutrino" field, the "electron" field, the "down quark" field, and the "up quark" field. The "disturbances" of these fields we accept as fundamental elementary particles: the electron neutrino (ν_e), the electron (e^-), the down quark (d), and the up quark (u). The most important properties of fundamental elementary particles are: the rest mass, the electric charge, and the spin.

The SI unit of mass is the kilogram (kg). However, for a fundamental elementary particle it is more convenient to use the non-SI unit of mass — the electronvolt divided by the speed of light squared (eV/c^2). An "electronvolt" is the energy acquired by an electron when it is accelerated by an electric potential difference of one volt. One eV/c^2 is approximately equal to 1.8×10^{-36} kg.

Measuring the electron neutrino mass is one of the great challenges of modern science. It is either extremely small or may even be exactly zero. The mass of an electron is approximately 0.51 MeV/c^2 (where M stands for "mega" $= 10^6$). Electrons and electron neutrinos together are called "leptons" (light particles). The masses of the up and down quarks are much greater than the electron mass. However, their values have not been measured directly because a free quark has never been observed and current theories predict that single quarks cannot be observed.

The SI unit of electric charge is the coulomb (C). It is convenient

5

to express the charges of fundamental elementary particles in units of a "fundamental charge" (e) which is equal to the magnitude of the electron charge, approximately 1.6×10^{-19} C. Unlike mass, all fundamental elementary particles have values of electric charge which are simple multiples of the fundamental charge. In units of the fundamental charge, the charge of an electron neutrino is zero, the charge of an electron is (-1), the charge of the up quark is $(2/3)$, and the charge of the down quark is $(-1/3)$.

The spin of a fundamental elementary particle can be thought of very roughly as analogous to the angular momentum of a rotating ball. The SI unit for angular momentum is the joule·second (Js). It is convenient to express the spin of a fundamental elementary particle in units of Planck's constant divided by 2π, which is approximately equal to 1.05×10^{-34} Js. (This unit is commonly designated $\hbar$.) In units of $\hbar$, all four particles of the first generation have the same spin, $1/2$. Note also that all particles of half-integer spin are called "fermions" after E. Fermi.

The "second generation" of the fundamental fields and their disturbances (the particles) is much more exotic than the first generation. It also consists of two leptons — the "muon neutrino" (ν_μ) with zero electric charge and the "muon" (μ^-) with electric charge (-1), and two quarks — the "strange quark" (s), with electric charge $(-1/3)$, and the "charm quark" (c), with electric charge $(2/3)$. The main difference between the two generations is the rest mass of the particles, which is much greater for the second generation. The only directly measured rest mass is that for the muon, approximately 106 MeV/c^2.

The even more exotic "third generation" consists again of two leptons — the "tau neutrino" (ν_τ) with zero electric charge and the "tau lepton" (τ^-) with electric charge (-1), and two quarks — the "bottom quark" (b) with electric charge $(-1/3)$ and the "top quark" (t) with electric charge $(2/3)$. The only measured mass is that of the tau lepton — approximately 1800 MeV/c^2.

The complete picture for quarks is a bit more complicated. There are three variants of each quark, called "blue" (B), "green" (G) and "red" (R) quarks. For example, there exist uB quarks, uG quarks and uR quarks, dB, dG and dR quarks, and so on. Thus, we have a total of 18 different quarks.

To complete the list of fundamental elementary particles (and the corresponding fields) we must add the "antiparticles". Each fundamental elementary particle has a corresponding antiparticle of the same mass but with some properties that are "opposite," e.g. electric charge. For example, the electron e^- has an electric charge (-1), and the "antielectron," which is also called the "positron," has an electric charge $(+1)$. That is why a positron is designated as e^+. Electrically neutral particles also have antiparticles. The electron neutrino, ν_e, does not have an electric charge, and an "electron antineutrino," $\bar{\nu}_e$, also does not have an electric charge.

Again, the situation with quarks is more complicated. An "antiquark" not only has the opposite electric charge but it also has "opposite color". As an example, an "up antiquark" $\bar{u}$ can be "antiblue" $(\overline{uB})$, "antigreen" $(\overline{uG})$ or "antired" $(\overline{uR})$. In the same way, there exist three "down antiquarks" $(\overline{dB})$, $(\overline{dG})$ and $(\overline{dR})$. The "color" of the quarks and antiquarks has nothing to do with the visual color of macroscopic objects but represents an intrinsic property like an electric charge. The electric charge of all up antiquarks is $(-2/3)$, and the electric charge of all down antiquarks is $(1/3)$ — opposite to the charges of the corresponding quarks. Note that all fundamental elementary particles have the same spin, $1/2$. Tables 1 and 2 summarize the properties of the fundamental elementary particles.

2.2. Interactions

According to the current scientific picture, the fundamental elementary particles described above do not interact with each other directly. There exist additional fields, called "gauge fields", which are responsible for the interactions between the fundamental elementary particles. The fundamental elementary particles interact with the gauge fields and, via these fields, with each other. There exist eight "gluon fields", an "electromagnetic field," and three "weak fields" which are responsible for the main interactions between fundamental elementary particles. There is also the "gravitational field" which appears to be much weaker and more complicated than all the other fields.

First generation

Name	Symbol	Rest mass (if known) in MeV/c²	Electric charge	Symbol for antiparticle
Electron	e^-	0.511	-1	e^+
Electron neutrino	ν_e		0	$\overline{\nu}_e$

Second generation

Muon	μ^-	105.7	-1	μ^+
Muon neutrino	ν_μ		0	$\overline{\nu}_\mu$

Third generation

Tau	τ^-	1777	-1	τ^+
Tau neutrino	ν_τ		0	$\overline{\nu}_\tau$

Table 1. Fundamental elementary particles: Leptons.

First generation

Name	Symbol	Electric charge	Symbol for antiparticle
Up	u	2/3	$\overline{u}$
Down	d	-1/3	$\overline{d}$

Second generation

Charmed	c	2/3	$\overline{c}$
Strange	s	-1/3	$\overline{s}$

Third generation

Top	t	2/3	$\overline{t}$
Bottom	b	-1/3	$\overline{b}$

Table 2. Fundamental elementary particles: Quarks.

The gluon fields are responsible for the "color interaction" between quarks. Due to this interaction, quarks of the same color, say red and red, repel each other, and the quarks of different colors, say blue and antiblue or red and green, attract each other. This attraction is so strong that all existing quarks are confined into compound particles, and a free quark has never been observed. That is why the quark–gluon interaction is also called the "strong nuclear interaction".

The electromagnetic field is responsible for interactions between all fundamental elementary particles having electric charge, i.e. all the fundamental elementary particles except the neutrinos. Two particles with the same sign of electric charge, i.e. positive and positive or negative and negative, repel each other, while the particles with opposite signs of electric charge attract each other. At small separations the electromagnetic interaction between quarks is much smaller than the strong nuclear interaction. But the range of the strong interaction is estimated to be quite short, about 10^{-15} m, while the range of the "electromagnetic interaction" is believed to be infinite. For electrons, muons and tau leptons, which do not interact with the gluon fields, the electromagnetic interaction is the strongest interaction.

All elementary particles interact with the weak fields which cause the "weak interaction" between particles. This interaction is much weaker than the electromagnetic interaction and has a range of about 10^{-18} m, i.e. 1000 times less than the strong nuclear interaction. That is why it is very difficult to detect a neutrino, it does not interact with either the gluon or the electromagnetic fields. On the other hand, the weak interaction causes many important transformations. The best known of them is the transformation of a down quark (d) into an up quark (u) with the creation of an electron (e^-) and an electron antineutrino ($\bar{\nu}_e$):

$$d \rightarrow u + e + \bar{\nu}_e. \tag{2.1}$$

All fundamental elementary particles interact with the "gravitational field." It causes an extremely small attractive "gravitational interaction" between the particles. Normally, the gravitational interaction between two fundamental elementary particles is negligible. However, its range, like the range of the electromagnetic interaction, is infinite and so it is very important to the dynamics of large electrically neutral objects like stars, planets and galaxies.

It would be wrong to think that the gauge fields are only passive mediators of interactions. They have their own disturbances which behave like elementary particles. The best known "gauge particle" is the photon (γ), a disturbance in the electromagnetic field. It has zero rest mass, zero electric charge, and its spin is 1. (Particles of integer spin are called "bosons," after S. Bose.) The three "weak gauge particles" W^+, W^- and Z^0, corresponding to the three weak fields are called the "intermediate vector bosons." The upper index indicates their electric charge ($+1$, -1, and 0). These particles have very large rest masses — approximately 80 GeV/c^2 for W^+ and W^-, and 91 GeV/c^2 for Z^0, (where G stands for "giga" $= 10^9$). The spin of all these particles is 1. The gauge particles of the eight gluon fields are called gluons, and the gauge particle of the gravitational field is called the graviton. Neither gluons or gravitons have been directly observed.

2.3. Compound "Elementary" Particles

Due to the powerful attraction between quarks of different colors, all existing quarks are confined into particles which are called "hadrons". As nobody is able to split hadrons into quarks, the hadrons are also considered to be elementary particles. There are two types of hadrons — "baryons" and "mesons."

A meson is a bound pair consisting of a quark and an antiquark of "opposite" colors. For example, a pion (π^+) consists of an up quark (u) and a down antiquark ($\bar{d}$) which can be red and antired, or green and antigreen, or blue and antiblue combinations:

$$\pi^+ = (uR, \overline{dR}) \quad \text{or} \quad (uG, \overline{dG}) \quad \text{or} \quad (uB, \overline{dB}). \qquad (2.2)$$

All these color combinations are identical because $\bar{R}R$, $G\bar{G}$ and $B\bar{B}$ cancel out forming a "colorless" pion which is observed in experiments. The electric charge of the pion is the algebraic sum of the electric charges of u and $\bar{d}$:

$$2/3 + 1/3 = 1.$$

The symbol "+" in π^+ indicates that the electric charge is $+1$ (in units of the fundamental charge). The spin of a pion is zero. The "antipair", $\bar{u}d$, i.e. $(\overline{uR}, dR)$ or $(\overline{uG}, dG)$ or $(\overline{uB}, dB)$ is designated

π^-. It has an electric charge (-1), spin zero and is considered the "antiparticle" of π^+. All mesons are colorless, have an integer electric charge $(0, +1, -1)$ and integer spin $(0$ or $1)$.

A baryon is a bound system consisting of three quarks of different colors. The most important of them — the "proton" (p) — consists of two up quarks (u) and one down quark (d), each of a different color. It may be any of the (uB, uG, dR) or (uB, uR, dG) or (uG, uR, dB) combinations. All three combinations are identical because the three different colors, G, R and B, cancel out, like color–anticolor, forming a colorless proton. The electric charge of the proton is,

$$2/3 + 2/3 - 1/3 = 1\,,$$

and it has spin $1/2$.

The second important baryon — the "neutron" (n) — consists of one up and two down quarks of different colors, (uB, dG, dR) or (uG, dB, dR) or (uR, dB, dG). All three combinations are identical, like the three combinations for the proton. The electric charge of a neutron is,

$$2/3 - 1/3 - 1/3 = 0\,,$$

and its spin is $1/2$.

Similarly, all other baryons are colorless combinations of three quarks. They have a spin $1/2$ and an integer electric charge. All "antibaryons" are colorless combinations of three antiquarks. For example, an "antiproton" $(\bar{p})$ consists of two up antiquarks $(\bar{u})$ and one down antiquark $(\bar{d})$. These antiquarks have different anticolors — one of them is antiblue, one is antigreen and one is antired. The spin of an antiproton is $1/2$, and its electric charge is (-1). Table 3 and Fig. 1 show the properties of a few hadrons.

2.4. Transformations of Elementary Particles

An amazing feature of the particle-field interactions is that these interactions cause not only attraction or repulsion between particles, but also the creation (production or emission) and destruction (decay or absorption) of particles. All these transformations are called

Name	Symbol	Composition	Rest mass in MeV/c²	Electric charge	Spin	Symbol for antiparticle
Proton	p	uud	938.3	1	1/2	$\bar{\text{p}}$
Neutron	n	udd	939.6	0	1/2	$\bar{\text{n}}$
Pion	π^+	u$\bar{\text{d}}$	140	1	0	π^-

Table 3. Selected hadrons.

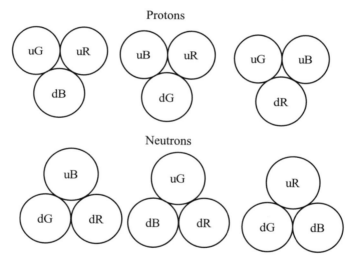

Fig. 1. Three combinations of up and down quarks form identical colorless protons or neutrons.

"reactions" in analogy with chemical reactions. The most common reaction is the spontaneous decay of a particle. There are only two known stable hadrons: the proton (p) and the antiproton ($\bar{p}$). All other hadrons and all the mesons decay into lighter particles soon after their creation. In all reactions the total energy, momentum, angular momentum and the electric charge are conserved. But the number of particles, and even their total rest mass is not conserved!

We mention here a few hadron decay reactions. A free neutron decays into a proton, electron and electron antineutrino:

$$n \rightarrow p + e^- + \bar{\nu}_e. \tag{2.3}$$

This reaction is the result of the decay of a down quark "hidden" inside the neutron. (See the reaction (2.1).) The mean lifetime of the neutron is about 15 minutes. Similarly, an antineutron decays into an antiproton, a positron and an electron neutrino:

$$\bar{n} \rightarrow \bar{p} + e^+ + \nu_e. \tag{2.4}$$

Comparing reactions (2.3) and (2.4) one can see that all particles in (2.3) are replaced by their antiparticles and vice versa. This is a general rule: in any formula for a reaction with elementary particles, one can make such a replacement to get a true new reaction.

The pion π^+ has a mean lifetime of only 26 nanoseconds (1 ns = 10^{-9} s), and it decays into an antimuon and a muon neutrino:

$$\pi^+ \rightarrow \mu^+ + \nu_\mu. \tag{2.5}$$

One can easily write the corresponding reaction for decay of antipion π^-.

The most important lepton — the electron — is stable, but the muon and the tau lepton are unstable having mean lifetimes of 2.2 microseconds and 290 femtoseconds, respectively (1 μs = 10^{-6} s, 1 fs = 10^{-15} s). The muon decays into three particles:

$$\mu^- \rightarrow e^- + \bar{\nu}_e + \nu_\mu. \tag{2.6}$$

The tau lepton has two major decay channels:

$$\tau^- \rightarrow \mu^- + \bar{\nu}_\mu + \nu_\tau, \tag{2.7}$$

$$\tau^- \rightarrow e^- + \bar{\nu}_e + \nu_\tau.$$

Perhaps the most impressive reaction is the annihilation of a particle and the corresponding antiparticle with the emission of photons. For example, if an electron e^- meets a positron e^+ (its antiparticle), they can annihilate, i.e. completely disappear and generate two photons of high energy:

$$e^- + e^+ \rightarrow 2\gamma. \tag{2.8}$$

The same reaction can be observed for a proton p and an antiproton $\bar{p}$:

$$p + \bar{p} \rightarrow 2\gamma. \tag{2.9}$$

Annihilation seems to be the only way to convert the total rest mass hidden in elementary particles into electromagnetic energy (photons).

The reaction which is the opposite of annihilation is "pair production," the creation of a "particle–antiparticle" pair. For example, if a neutron n absorbs a photon γ, an up quark (u) up antiquark ($\bar{u}$) pair can be created inside the neutron:

$$\gamma + n = \gamma + (udd) \rightarrow udd u\bar{u}. \tag{2.10}$$

This combination decays into a proton p and an antipion π^-:

$$udd u\bar{u} \rightarrow (uud) + (\bar{u}d) = p + \pi^-. \tag{2.11}$$

(We do not show all the possible colors — all composite particles are colorless.)

All possible reactions in the world of elementary particles are restricted by the conservation laws. Besides the four classical laws — conservation of energy, momentum, angular momentum, and electric charge — additional empirical conservation laws have been discovered. The first of these laws, the "Law of Conservation of Baryon Number" states that the difference between the number of baryons and antibaryons remains constant. For example, in the reaction (2.3) we have one baryon (neutron) before the reaction, and one baryon (proton) after the reaction. In the reaction (2.9) we have one baryon p and one antibaryon $\bar{p}$ (the difference is $1 - 1 = 0$) on the left side, and no baryons on the right side. The second empirical law deals with the leptons: the "Law of Conservation of Lepton Number". The difference between the number of leptons and the number of antileptons is conserved. Moreover, this difference is separately conserved for each generation! As an example, in reaction (2.3) there are no leptons before the reaction, and there are a lepton e^- and an antilepton $\bar{\nu}_e$ after the reaction (the difference is $1 - 1 = 0$).

2.5. Accelerators

The best way to study the interactions between fundamental elementary particles is to make them collide with each other. The greater the range of available collision energies for the particles, the more interesting the effects that can be observed. The same is true for the compound elementary particles: studying their collisions is the only way to understand their internal structure. And, finally, to create a new particle whose rest mass is greater than the rest mass of the colliding particles, we have to increase the energy of the colliding particles.

To gain this energy scientists use accelerators — giant factories with highly sophisticated equipment. There are two basic designs of accelerators. The first one is the "linear accelerator" or "linac" where particles move along a straight line, gradually increasing their speed. The linac consists of a set of metal tubes separated by gaps. A particle with electric charge is accelerated by electric fields produced in the gaps. The longest linac, the Stanford Linear Accelerator (SLAC) in California, has a length of approximately 3 km and can accelerate electrons and positrons to energies of 50 GeV. The second basic design of accelerators is the "synchrotron." In a synchrotron, particles move along a circular path under the action of a magnetic field, produced by powerful electromagnets. When particles pass through regions of applied electric fields, they gain additional energy. The electric and magnetic fields must be strictly synchronized with the motion of the particles to keep them moving on the correct circular path. The most powerful synchrotron, the Tevatron, at Fermilab in Illinois has a circumference of approximately 6 km. It accelerates protons and antiprotons to energies of 1 TeV (T stands for tera, which means 10^{12}). The largest synchrotron, the Large Electron Positron Collider (LEP) at CERN has a circumference of approximately 27 km. (CERN stands for research center of the European Council for Nuclear Research, and LEP is located on the border between France and Switzerland, 100 m beneath the ground.) In the LEP, electrons and positrons can collide at an energy of up to 180 GeV. Currently under construction at CERN is the Large Hadron Collider (LHC) which will provide collisions between protons and antiprotons with a total energy of 14 TeV. There are many other well-known accelerators around the globe. As an example, we mention the Relativistic Heavy Ion Collider (RHIC) at Brookhaven National Laboratory in New York. Its synchrotron has a 3.8 km circumference, and can accelerate protons and other atomic nuclei from deuterium to gold.

Using accelerators, scientists hope to discover new fundamental elementary particles having rest masses of the order or greater than the mass of a top quark which is estimated to be 175 GeV/c^2. The "most wanted" particle is probably the Higgs boson, named after P. Higgs. Unlike all other fundamental elementary particles, the theoretically predicted Higgs boson does not have a spin and is of great importance in understanding how and why other particles acquire their masses.

There are many other important phenomena which can be observed using accelerators. As an example, heavy nuclei colliding at high energy produce a short-lived "fireball" of quarks and gluons, called a "quark-gluon plasma." Such collisions are of great current interest.

2.6.   Nuclei

While protons and neutrons are "colorless" particles, the gluon fields provide strong nuclear interaction between two protons, two neutrons or between a proton and a neutron. A quark from one proton (or neutron) inside a nucleus can often interact "slightly" with a quark from a different nearby proton (or neutron). These interactions are attractive on average and at a distance of a few femtometers (1 fm = 10^{-15} m) this "residual interaction" exceeds the strength of the electromagnetic interaction (repulsion) between two protons. The distance between a proton and a neutron in a nucleus is of the order of 1 fm, and the residual strong interaction is estimated to be up to 20 times greater than the electromagnetic interaction.

Due to this residual strong interaction, protons and neutrons can attract each other, forming stable atomic nuclei. Nuclei with the same number of protons but different number of neutrons are called "isotopes" of the same chemical element. Nuclei with only one proton are called hydrogen (H) nuclei. Hydrogen has two stable isotopes. The nucleus of the first one is a single proton, p, which is designated as ^1_1H. (The subscript shows the number of protons (atomic number), and the superscript shows the total number of protons and neutrons (mass number).) The nucleus of the second stable isotope of hydrogen, ^2_1H, deuterium contains one proton and one neutron. This nucleus has a special name, the "deuteron". Deuterons can be created when neutrons are captured by protons:

$$n + {}^1_1H \rightarrow {}^2_1H + \gamma. \tag{2.12}$$

Only 0.015% of all hydrogen nuclei in nature are deuterons. This number, 0.015%, is called the "natural abundance" (or just abundance) of the ^2_1H isotope.

The nuclei of all isotopes of the same element have the same electric charge but, generally, different spin (the intrinsic angular momentum). For example, the spin of the proton, ^1_1H, is 1/2 (in units of the Planck constant, $\hbar$). The spin of the deuteron, ^2_1H, is 1. Nuclei with an even number of protons and an even number of neutrons do not have spin. All nuclei which have a spin behave like tiny "permanent magnets". Like all permanent magnets they produce a magnetic field. The magnitude of this magnetic field is proportional to the quantity called the "intrinsic magnetic moment" of the nucleus. The unit of magnetic moment in the SI system is Am^2 (ampere meter squared) or J/T (joule/tesla). 1 Am^2 is the magnetic moment of a current loop of area 1 m^2 carrying an electric current of 1 A. The intrinsic magnetic moments of nuclei are normally measured in units of the "nuclear magneton," $\mu_n \approx 5.05 \times 10^{-27}$ J/T. The magnetic moment of a proton in these units is 2.79 μ_n. The magnetic moment of a neutron is 1.9 μ_n and the magnetic moment of a deuteron is 0.86 μ_n.

The nuclei which contain two protons are called helium (He) nuclei. There are two stable isotopes of helium, ^3_2He and ^4_2He. The first one has a spin 1/2. Its natural abundance is only 10^{-4}%. The "main" isotope, ^4_2He, does not have a spin, or a magnetic moment.

The radius of nuclei (in fm) can be estimated approximately as 1.5 $A^{1/3}$, where A is the mass number, if the nuclei are assumed to have an approximately spherical form. The volume of nuclei is proportional to the mass number, so their density is approximately constant. It is estimated to be about 10^{17} kg/m^3. For stable nuclei with mass number up to 40, the number of neutrons is typically close (or equal) to the number of protons. For heavy nuclei, the number of neutrons is typically significantly greater than the number of protons. The heaviest stable nucleus is Bismuth, $^{209}_{83}Bi$ (83 protons and 126 neutrons). It has a spin of 9/2, an intrinsic magnetic moment of 4 μ_n, and 100% natural abundance (i.e. no other stable isotopes of Bi exist).

2.7. Nuclear Reactions

The most typical nuclear reaction is a nuclear decay. The majority of nuclei known to scientists are radioactive. Such nuclei are unstable: they quickly or gradually decay into more stable nuclei by emitting particles, electrons (e^-), positrons (e^+), photons (γ), electron neutrinos (ν_e), electron antineutrinos $(\bar{\nu}_e)$, protons (p), neutrons (n), helium nuclei (^4_2He) or even splitting into smaller pieces ("fission"). The lightest radioactive nucleus is the hydrogen isotope ^3_1H which is called "tritium." It decays into ^3_2He by emitting an electron and an electron antineutrino:

$$^3_1H \rightarrow {}^3_2He + e^- + \bar{\nu}_e. \tag{2.13}$$

The emitted electron is called a "β^--particle." The decay itself is called a β^--decay. β^--decay is associated with the weak interaction which causes the transformation of a neutron into a proton inside a nucleus. (See reaction (2.3).) An initial amount of tritium nuclei decreases by half in approximately 12 years. This characteristic time is called the "halflife" of the nucleus.

Many radioactive nuclei exhibit β^--decay. An extremely important example is the β^--decay of the carbon isotope, $^{14}_6C$, into the nitrogen isotope $^{14}_7N$:

$$^{14}_6C \rightarrow {}^{14}_7N + e^- + \bar{\nu}_e. \tag{2.14}$$

$^{14}_6C$ nuclei are constantly being produced in the atmosphere, due to reactions with cosmic rays which strike the atmosphere from space. As a result, approximately 1.3×10^{-12} of all carbon nuclei in the atmosphere are $^{14}_{6}C$ nuclei. Living organisms constantly take in and release CO_2 to and from the air. Thus, 1.3×10^{-12} of carbon nuclei in living organisms are also $^{14}_{6}C$ nuclei. When an organism dies, it stops exchanging carbon with the air. New $^{14}_{6}C$ nuclei do not come into the organism, and the old ones gradually decay away. (The halflife of $^{14}_{6}C$ is approximately 5730 years.) Thus, by measuring the relative amount of $^{14}_{6}C$ nuclei a scientist can find out when an organism died. This is the basis for "carbon dating" many artifacts made from organic materials.

If a proton inside a nucleus transforms into a neutron, the nucleus emits a positron and an electron neutrino. As an example, the nitrogen isotope $^{13}_{7}N$ decays into the carbon isotope $^{13}_{6}C$, emitting a positron:

$$^{13}_{7}N \rightarrow {}^{13}_{6}C + e^+ + \nu_e \,. \tag{2.15}$$

Such a decay is called a β^+-decay and the emitted positrons are called β^+-particles. It is also caused by the weak interaction.

The third type of decay caused by the weak interaction is called electron capture β-decay: a proton inside a nucleus absorbs an electron and transforms into a neutron with the emission of an electron neutrino. The lightest radioactive isotope which experiences electron capture is $^{7}_{4}Be$:

$$^{7}_{4}Be + e^- \rightarrow {}^{7}_{3}Li + \nu_e \,. \tag{2.16}$$

Electron capture β-decay is possible only if there is an electron somewhere around the nucleus. There thus exists an exciting opportunity to manipulate the rate of the electron capture β-decay by manipulating the electron which is to be captured.

Another important type of nuclear decay is connected with the emission of $^{4}_{2}He$ nuclei which are called "α-particles." The corresponding decay is called "α-decay." The lightest nucleus which experiences α-decay is $^{5}_{2}He$:

$$^{5}_{2}He \rightarrow {}^{4}_{2}He + n \,. \tag{2.17}$$

This reaction may also be considered as neutron emission.

All nuclei with atomic number greater than 83 are unstable to α-decay due to the electromagnetic interaction (electric repulsion) between protons. The heaviest long-lived nucleus found in nature is the uranium isotope $^{238}_{92}U$ with a halflife of approximately 4.5×10^9 years. It experiences α-decay into the thorium isotope $^{234}_{90}Th$:

$$^{238}_{92}U \rightarrow {}^{234}_{90}Th + {}^4_2He. \qquad (2.18)$$

Nuclear decay is often followed by the emission of a photon or a neutron. The emission of a photon by a nucleus is called "γ-radiation." The photons emitted by nuclei normally have very high energy, often a million times greater than the energy of the photons of visible light.

For a long time, scientists have tried to study the "transuranic elements": nuclei with atomic number greater than 92. Two of them, neptunium $^{237}_{93}Np$ and plutonium $^{244}_{94}Pu$, have been found in nature in very small quantities. To produce heavier nuclei scientists bombard target nuclei with ion beams of lighter or heavier nuclei. In this way they have produced nuclei with atomic number up to 118. Normally, these artificial nuclei are extremely unstable: their halflives are much less than a second. However, there exist relatively stable isotopes even among the transuranic nuclei: one of them, $^{289}_{114}X$ (X means that the name is not given yet), has a halflife of approximately 30 seconds.

The transuranic elements experience spontaneous fission: the decay of a nucleus into two approximately equal-sized fragments. For uranium $^{238}_{92}U$ the probability of fission is approximately a million times smaller than the probability of α-decay. With increasing atomic number, the probability of fission quickly increases. Fission induced by neutrons is used in nuclear power plants and engines.

Perhaps the most important nuclear reaction is "fusion", which is responsible for the energy production in stars. For example, two hydrogen nuclei ^1_1H (protons) can fuse producing deuterium ^2_1H:

$$^1_1H + {}^1_1H \rightarrow {}^2_1H + e^+ + \nu_e. \qquad (2.19)$$

Deuterium ^2_1H and tritium ^3_1H can combine to produce a helium nucleus ^4_2He :

$$^2_1H + {}^3_1H \rightarrow {}^4_2He + n, \qquad (2.20)$$

and so on.

Scientists believe that fusion can be used as a source of energy on Earth. It could become an inexpensive and clean energy source which produces plenty of energy and a small amount of short lived radioactive waste elements. To use fusion, one must have a fuel at very high temperature so as to overcome the repulsive electromagnetic interaction between the nuclei. If nuclei can approach each other closely enough, the strong nuclear interaction comes into play. At a distance of the order of 10^{-15} m the strong interaction becomes greater than the electromagnetic interaction. The two nuclei combine into a heavier nucleus producing a lot of energy.

The first problem on the long way to controlled fusion is the heating of the fuel to a temperature of the order of $10^8\,°C$. When the temperature of the fuel increases, the electrons are stripped from the atoms. For example, for deuterium–tritium fuel one will get an overall electrically neutral gas consisting not of atoms or molecules but from independently moving electrons and nuclei, e^-, ^2_1H and ^3_1H. Such a highly ionized gas is called a "plasma." Plasmas conduct electric currents and they can be used to further increase the temperature of a plasma. Electric currents, microwave and laser radiation, even beams of high energy particles — all have been used to further heat plasmas.

The most important problem is how to prevent the plasma from cooling and to maintain temperatures of about $10^8\,°C$. The most natural way is to trap the plasma with a magnetic field, which can keep the electrically charged particles confined inside a small volume of space. There exist many experimental stations around the globe which are used to implement ideas for a fusion power plant. Probably the most famous of them is the Joint European Torus (JET) in England which can already generate small amounts of fusion power.

2.8.   Atoms — Basic Units of Matter

Due to the electromagnetic interaction, nuclei and electrons attract each other and form bound neutral atoms. In a neutral atom the number of electrons is equal to the number of protons in the nucleus. Thus, the total electric charge of an atom is zero.

The internal properties of an atom differ drastically from the familiar properties of macroscopic particles. The internal energies of an atom take on a discrete set of values. For example, the hydrogen atom which consists of the hydrogen nucleus (proton) and an electron can have a minimum value of its internal energy of approximately -13.6 eV. The next possible value of its internal energy is -3.4 eV and so on. If one measures the internal energy of a hydrogen atom, one will never get the value -10 eV or -5 eV. Normally an atom is found in the ground state — the state of minimum energy. The states corresponding to other energies are called excited states. Together the ground state and the excited states form the set of "stationary states" — the states with definite energies. Using an electric current, or beams of particles, or beams of light, one can transfer atoms from the ground state into the excited states. Normally the lifetime of an atomic excited state is very short — much smaller than a second ($\approx 10^{-8}$ s!). Due to the electromagnetic interaction, an atom "jumps" (makes a transition) from an excited state to the ground state (or into an intermediate excited state and then into the ground state) by emitting a photon (or photons). There also exist "metastable" atomic states which do not decay for a relatively long time (from milliseconds to minutes).

The stationary states are only a tiny part of all possible states of an atom. An atom can be in a superposition of two or more stationary states. This means that an atom can be in two or more states of different energies at the same time! It is difficult for us to get a feel for it but we have to accept this fact, because it is based on the results of a large number of experiments. As an example, an atom can be in an equal superposition of two states of energy, E_1 and E_2. If one measures the atom's energy, one will get with equal probabilities E_1 or E_2. After the measurement the atom has a definite energy, but before the measurement it did not. Such "superposition states" are of great current interest in physics.

An electron in an atom does not really appear as a point-like particle orbiting a point-like nucleus, but rather as a big negatively charged cloud surrounding a tiny positively charged cloud (nucleus). For different stationary states (i.e. the states with definite energy) the "electron cloud" has different shapes. For a superposition of stationary states the shape of the electron cloud continuously changes with time.

If an atom has more than one electron, the electron clouds largely overlap. It is then impossible to distinguish the two or more electrons inside the same atom. An atom has a common electron cloud whose electric charge is equal to the sum of electric charges of all atomic electrons. Atoms can also combine in various ways to form "molecules", new compound substances. While atoms and molecules are electrically neutral (i.e. the total electric charge is zero), there still are residual electric interactions between them. As a result, atoms and molecules can condense into many familiar liquid and solid substances.

2.9. Quantum Motion and Schrördinger Cat States

We have already mentioned that an atom does not have to be in a state with a definite energy: it can be in a state which is an arbitrary superposition of stationary states, each with a definite energy. Nevertheless, one normally finds an atom in one of its stationary states: the ground state, the first excited state, the second excited state, and so on. The first question that naturally arises is: why a superposition e.g. of the ground state and the first excited state is less likely than either of these two states individually? The answer is connected to interactions with the "environment." The environment includes all the quantum fields in space (whose disturbances we observe as particles) as well as particles themselves. Due to the interactions with the environment, the lifetime of any superposition state is usually less than the lifetime of a stationary excited state.

The next question that arises is: does this property remain if we consider the very high energy levels of an atom. The surprising answer is: no. In this case, the most stable states are not stationary states but rather so-called "quasi-classical states." A quasi-classical state is a superposition of a large number of stationary states which describes a very small, point-like electron cloud. Thus, in the limit of very highly excited atomic states we transfer to the familiar classical picture of electron motion.

So far we have discussed the interior properties of an atom. What about the motion of an atom as a single particle? If we consider the motion of an atom as a single particle, we typically face the obvious fact: the kinetic energy of a moving atom is much larger than its minimum possible kinetic energy. Thus, an atom as a whole unit is usually in a quasi-classical state. It therefore travels like a familiar macroscopic particle.

A most intriguing question now appears: can one produce a superposition of quasi-classical states, e.g. the same atom in two macroscopically different positions at the same time? It looks like the answer is positive. It is difficult to imagine this, but we have to accept it. Under special circumstances an atom can be transformed into two "quasi-atoms" which travel in two macroscopically different directions. Both quasi-atoms have the same size and structure as the initial atom. Each quasi-atom travels in the same way as if there is no second quasi-atom at all.

And what about macroscopic particles which consist of thousands of atoms? Can such a particle be transformed into two "quasi-particles" which have the same size and shape as the initial particle? To sharpen this question Schrödinger famously considered a cat which became a superposition of two "quasi-cats", one of them alive and the other one dead. Because of this, superposition states involving macroscopic objects or even macroscopically distinguishable trajectories of an atom are called Schrödinger cat states. Incredibly, physics gave a positive answer to the question about macroscopic particles! Recently, Schrödinger cat states were brought under experimental investigation.

The next question that arises: why don't we normally observe such fantastic states in the macroscopic world? The answer again is connected to interactions with the environment. Due to these interactions, the Schrödinger cat state normally decays very quickly. The greater the mass of a particle, the smaller the lifetime of its Schrödinger cat states. The most mysterious and so far unanswered question is the following: how do the Schrödinger cat states decay? If we have two quasi-particles originated from the same particle, they quickly become part of the environment, and they "entangle" with the environment. This process is called "decoherence". Then, the environment somehow destroys one of the quasi-particles and transforms the other one into the real particle. The most surprising fact is that this quantum transformation occurs instantaneously even for two quasi-particles which are separated by a significant distance! This very curious property disturbed many physicists. The next question is: what happens if we try to "catch" one of the quasi-particles? The answer is also known. If we try to do this, we artificially strengthen the interaction with the environment. The Schrödinger cat state immediately decays, and we catch a real particle in one of two possible locations and nothing remains in the other one.

Chapter 3

APPLICATIONS OF PHYSICS

3.1. Magnetic Resonance

In addition to its electric properties, an electron also behaves like a tiny magnet. To describe the magnetic properties of particles scientists use the magnetic moment. The same quantity is also used to describe the magnetic properties of a magnetic needle or an electric current loop. The magnetic moment of an electron is very small: it is approximately equal to 9.3×10^{-24} Am^2. This quantity is called the "Bohr magneton," μ_B, after N. Bohr. Note that we describe here the interior intrinsic magnetic moment of an electron which is unconnected to the orbital motion of the electron. Because of the electron's charge, an orbital motion "generates" an additional orbital magnetic moment, just like a current in a macroscopic current loop.

The intrinsic magnetic moment of the electron is closely connected to its intrinsic angular momentum, i.e. its spin. The magnetic moment is a vector quantity, and it is always antiparallel to the electron's spin. The ratio of the magnetic moment to the spin is often called the gyromagnetic ratio. For the electron, the gyromagnetic ratio is approximately equal to the ratio of its electric charge to its mass.

The electronic magnetic moment of an atom can be considered to be the vector sum of the magnetic moments of the individual electrons. If the total electronic magnetic moment is zero, the atom is "diamagnetic." Otherwise the atom is "paramagnetic." The same terms are also applied to ions. The smallest non-zero electronic magnetic moment of an atom is equal to the intrinsic magnetic moment of one electron. If such an atom is placed into an external magnetic field, its ground state (the state with minimum energy) splits into two stationary states: one of them corresponds to the magnetic moment oriented in the direction of the magnetic field, the other one corresponds to the opposite orientation of the magnetic moment.

The energy difference between these two states is proportional to the strength of the magnetic field. The unit of the magnetic field is the tesla (T). For example, the Earth's magnetic field is approximately 50 μT (1 $\mu T = 10^{-6}$ T). In a magnetic field of 1 T, the magnetic energy difference between the two states is about 2×10^{-23} joules (J). If we divide this quantity by Planck's constant $h = 6.6 \times 10^{-34}$ Js, we get the frequency of the electromagnetic wave whose photon has an energy of 2×10^{-23} J. This frequency is about 30 gigahertz (1 GHz $= 10^9$ Hz).

If a sample containing many of the paramagnetic atoms considered above is placed into a magnetic field of 1 T, microwaves of frequency about 30 GHz will be absorbed by the sample. The atoms which absorb photons increase their magnetic energy, which is eventually converted into thermal energy. This phenomenon is called electron spin resonance (ESR) or electron paramagnetic resonance (EPR).

There exist a lot of modifications of the basic ESR experiment. One of the most impressive modifications is called electron spin echo. Spin echo can be observed when a sample containing paramagnetic atoms or ions is irradiated by short pulses of electromagnetic radiation of the ESR frequency. After the action of such resonant pulses, the sample emits its own electromagnetic pulse. (See Fig. 2.) This pulse is called the electron spin echo. Its frequency is also equal to the ESR frequency. With well defined pulse conditions, the shape of the spin echo reproduces the shape of the first electromagnetic pulse, justifying its name "echo". Using spin echo scientists can study interactions between magnetic moments and the transformation of magnetic energy into thermal energy of a sample.

So far we have discussed the magnetic moment of the electron. The majority of atomic nuclei also have a spin and an intrinsic magnetic moment. The magnetic moment of a nucleus is much smaller than the magnetic moment of an electron. Typically it is expressed in terms of the nuclear magneton $\mu_n \approx 5 \times 10^{-27}$ Am^2. As an example, the magnetic moment of a proton is approximately 2.8 μ_n. Even for the simplest nucleus — the proton — there is no simple expression

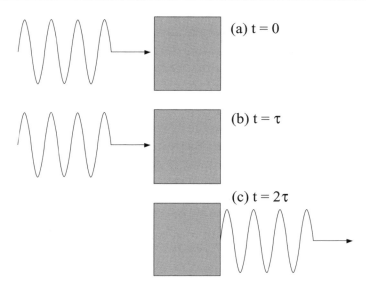

Fig. 2. Generation of a spin echo. (a) A sample (shown as a rectangle) is irradiated by the first electromagnetic pulse at time $t = 0$. (b) The sample is irradiated by the second pulse at time $t = \tau$. (c) At time $t = 2\tau$ the sample emits an electromagnetic pulse, which is called the "spin echo".

for the gyromagnetic ratio. If a diamagnetic atom, ion or molecule containing a nucleus with non-zero magnetic moment is placed in an external magnetic field, its ground state splits into two or more stationary states depending on the magnitude of the nuclear spin. As an example, the ground state of a diamagnetic molecule containing a hydrogen atom splits into two stationary states connected with the two possible orientations of the proton magnetic moment, just like in the case of the electron magnetic moment discussed above. The energy difference between the two states is much smaller than for the electron magnetic moment: for a proton in a magnetic field of 1 T it is only 2.8×10^{-26} J. The corresponding frequency of the nuclear magnetic resonance (NMR) is approximately 40 MHz (1 MHz $= 10^6$ Hz). Just as with electrons, magnetic nuclei can also produce spin echoes. While the ESR frequency normally belongs to the microwave region of the electromagnetic spectrum, the NMR frequency typically belongs to the FM radio wave region.

NMR and nuclear spin echo have found enormous applications in science and technology. The NMR frequency is very sensitive to both the external magnetic field and the structure and environment of a molecule. Thus, NMR can be used for the precise measurement of an external magnetic field and for the study of the molecular structure. Using a sophisticated sequence of electromagnetic pulses and computer analysis of the nuclear spin echo signals one can get information about the motion of proton magnetic moments in a small region of a human body. The output of this technique is well-known to the general public as magnetic resonance imaging (MRI).

Both ESR and NMR can be observed in a sample containing a macroscopic number of particles. Even in MRI, where the NMR signal comes from a small region of a human body, this small region still contains a macroscopic number of protons. The great challenge of magnetic resonance is to detect it from just a few atoms, and ultimately from a single atom in a sample.

How might one hope to achieve this goal? One of the approaches is called "magnetic resonance force microscopy (MRFM)." To detect ESR or NMR, the MRFM utilizes a very sensitive cantilever — a tiny bar, invisible to the human eye. One end of the cantilever is fixed. The other end can vibrate freely. (See Fig. 3.) The cantilever, like a swing, has its own resonant frequency. When the frequency of the spin motion matches the cantilever frequency, the cantilever starts to vibrate sympathetically. Even extremely small vibrations of a cantilever (of the order of an atomic radius!) can be detected by the optical methods, and so MRFM is a very sensitive technique.

Another approach to detecting magnetic resonance utilizes a "scanning tunneling microscope." In this remarkable device, a tungsten probe which has a very sharp tip, sometimes only one atom wide, is placed very close to a conducting surface. Electrons from the conducting surface can jump to the probe providing a small "tunneling" current, typically of a few picoamperes (1 pA $= 10^{-12}$ A). This current is very sensitive to the size of the gap between the probe and the surface: it can detect a change in the gap of the order of 100 picometers, i.e. of the order of atomic radius. Thus, moving the probe along the conducting surface, a scientist can trace the contours of the surface atoms or ions with magnification up to 10^8.

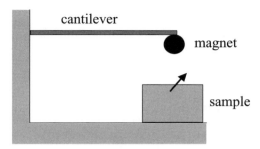

Fig. 3. Magnetic Resonance Force Microscopy (MRFM). A sample (shown as a rectangle) contains a paramagnetic atom. An arrow represents the magnetic moment of the atom. A small magnet is attached to the cantilever tip. The magnet interacts with the magnetic moment of the atom. This interaction causes the properties of the cantilever vibration to change, allowing the detection of the atom.

If a conducting surface contains paramagnetic atoms, and the whole sample is placed in the external magnetic field, scientists have observed oscillations (modulations) of the tunneling current. The modulation frequency matches the ESR frequency for the paramagnetic atoms. Because the tunneling current comes from the vicinity of a single atom, such experiments may result in a method for detecting ESR from a single atom.

3.2. Nanotechnology

The building blocks of modern electronics are rapidly shrinking, and research devices are now approaching the atomic scale. Feature sizes of the order of a few nanometers (1 nm $= 10^{-9}$ m) are becoming routine in such research devices. This area of scientific research is often referred to as nanotechnology. It includes many exciting research directions. We will mention only one of them, "nanotubes."

Probably everybody knows two forms of solid carbon: graphite and diamond. Today scientists work with new solid forms of carbon. One of them consists of big molecules which look like tiny cylinders. Such molecules are called nanotubes. This new form of carbon may withstand high stresses and temperatures. The diameter of nanotubes ranges from one to a few nanometers. The length of nanotubes can exceed one micrometer (10^{-6} m). Some nanotubes consist of many coaxial hollow cylinders. Other nanotubes consist of only one cylinder made of one layer of carbon atoms. Nanotubes may become the building blocks of future electronic devices instead of the silicon used in contemporary devices. (The smallest commercial conventional silicon devices have feature sizes no smaller than about 100 nm.)

There are a few different ways to produce nanotubes. For example, scientists may heat a piece of graphite with powerful laser pulses, producing carbon vapor which condenses and forms nanotube molecules. These molecules can then be collected on a surface, such as copper.

What makes nanotubes so promising a material for electronics? First of all, scientists can manipulate the electrical properties of nanotubes by changing their diameters and the number of coaxial cylinders in the tubes. Tubes can be designed to be very good electric conductors as well as semiconductors. In a semiconductor, electrons cannot move through a sample unless additional energy (e.g. light energy) is transferred to the electrons to put them into a conducting state. By increasing the diameter of the nanotubes scientists can smoothly decrease the amount of energy required to put electrons into a conducting state. It is estimated that good conducting nanotubes can withstand an electric current density of up to 10 teraamperes per meter squared (10^{13} A/m^2), one thousand times greater than the limit for copper.

3.3.   Superconductors

The phenomenon of superconductivity has been known for a long time, but it is still one of the hottest topics in science. To keep an electric current in an electric circuit, one needs an electromotive force (EMF), e.g. a battery. Without the EMF, the current quickly stops because of the finite electric resistance of the conductor. An electric current in a conductor is an ordered flow of conducting electrons. The electric resistance originates from the scattering of electrons by thermal vibrations of the positively charged ions or impurities in the conductor. In a typical superconducting material conducting electrons combine into pairs called "Cooper pairs" after L. Cooper. Unlike an electron which has a spin of 1/2, a Cooper pair has zero spin and is a boson. Bosons have an astonishing property: all of them can be in the same quantum state. So, many Cooper pairs form one big Cooper system which spreads out throughout the whole volume of the superconductor. When such a "monster" circulates in a superconducting loop, nothing can stop it.

The Cooper pairs normally appear at very low temperatures. That is why for a long time superconducting properties have been observed only at very low temperatures. Such temperatures are usually quoted in kelvin (K), the number of $°C$ above absolute zero $(-273.15°C)$ where all thermal motions cease. For example, lead becomes a superconducting material at approximately 7 K, mercury does at 4 K, and so on. Today scientists and engineers study complicated ceramic materials which become superconducting at temperatures of more than 100 K. As an example, copper oxide containing mercury, barium and calcium becomes superconducting at approximately 130 K.

Very impressive superconducting phenomena can be observed when two superconducting wires are separated by a thin layer of insulator. Such a combination is called a Josephson junction, after B. Josephson. The typical width of the insulating film is of the order of 1 nanometer $(10^{-9}$ m$)$. If a Josephson junction is inserted into an electric circuit, and the electric current in the circuit is less than some critical value, the Josephson junction does not show any electrical resistance, just like a pure superconductor (see Fig. 4). If the current in the circuit exceeds the critical value, the Josephson junction acquires an electrical resistance. It happens because an alternating current is produced in the junction, and it begins to radiate electromagnetic waves.

The Superconducting Quantum Interference Device ("SQUID") is a superconducting ring containing two Josephson junctions. When such a ring is inserted into an electric circuit (see Fig. 5) it does not show electric resistance until the electric current in the circuit becomes greater than some critical value. Because of "interference" between the two junctions, the value of the critical current shows an oscillating dependence on the magnetic flux through the superconducting ring. Thus, SQUIDs are routinely used for measurement of

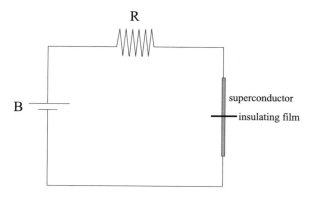

Fig. 4.   A Josephson junction in an electric circuit. B and R are the battery and resistor, respectively.

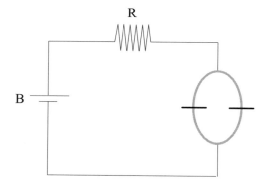

Fig. 5.   A superconducting loop with two Josephson junctions inserted into an electric circuit. The notation for the components is the same as in Fig. 4.

very weak magnetic fields, in particular, the magnetic fields produced by human tissues.

Because Cooper pairs are all in the same quantum state, scientists study the opportunity to create unusual superposition states for a macroscopic group of Cooper pairs. As an example, they experiment with Josephson junctions in which one of the superconductors is replaced by a small nanometer-scale "superconducting box." (See Fig. 6.) Using an electric field scientists can change the number of Cooper pairs in the superconducting box one at a time. Next, they produce a superposition of two macroscopic states of the box: one of

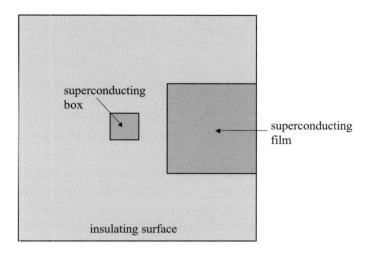

Fig. 6.   A Josephson junction on an insulating surface.

these states corresponds to N Cooper pairs in the box, the other one corresponds to $N + 1$ pairs, where N is a macroscopic integer.

Even more impressive experiments have demonstrated Schrödinger cat states for a superconducting current: the current flows in two opposite directions at the same time. It does not mean that some of the Cooper pairs flow in one direction, and the rest flow in the opposite direction: all the Cooper pairs flow in the two opposite directions at the same time!

The magnetic properties of a superconductor are also extremely interesting. It was believed for a long time that a superconductor can only generate a magnetic moment by electric current flow, i.e. without an electric current the magnetic moment of a superconductor is zero. For almost all known superconductors this statement is true. When a sample of such a typical superconductor is placed in a magnetic field, the magnetic field induces an electric current which flows without resistance along the surface of the sample. This current generates a magnetic moment which points in the direction opposite to the direction of the external applied magnetic field. Thus, if one approaches a piece of a typical superconductor with a pole of a magnet, the induced effect is that the magnet and the superconductor always repel each other. This fact is responsible for the well-known phenomenon of "magnetic levitation:" a small magnet can levitate above a superconductor cup due to this magnetic repulsion.

Currently, it is known that there exist non-typical magnetic superconductors which have an internal magnetic moment without any electric current, i.e. superconductor magnets! Some of these materials become superconducting at relatively high temperatures, above 50 K.

3.4.   Tsunami

The wide area of applied physics includes not only artificial designs from nanotubes to thermonuclear generators but also powerful natural phenomena. Tsunami is one such phenomenon. A wall of water up to 30 m in height and 30 km wide crashing on the shore at a speed of 15 m/s or more may bring death and massive destruction to coastal communities around the globe. Normally the waves in the oceans are generated by the wind. The length of these waves is relatively short, and the wave motion involves a relatively thin layer of water near the surface of the ocean. The height of the waves can reach up to 30 m but the water far below the surface remains undisturbed.

Typically, tsunami are generated by underwater earthquakes. They can propagate thousands of kilometers through the oceans with great speeds, up to 200 m/s. With a wavelength of up to 800 km and a height of a few meters tsunami remains almost invisible in the deep ocean. However this wave involves all the water from the surface to the bottom of the ocean. Near the coast, as the depth of the water decreases, the wavelength and speed of the wave decrease while the amplitude increases significantly.

In spite of the obvious difficulties, scientists are trying to create a reliable system of tsunami warning. The system which is known as the Deep-Ocean Assessment and Reporting of Tsunami (DART) is based on a network of deep-ocean stations. The basic idea is to measure pressure variations near the ocean bottom. While the value of the pressure at the bottom of the ocean is not affected by wind-generated waves, it is sensitive to tsunami which involve the whole body of water from the surface to the bottom. Information about tsunami must be converted into sound waves for transmission from the measurement device on the bottom to a buoy on the surface which, in turn, will relay the information to a shore ground station via a satellite using electromagnetic waves. The successful implementation of the DART project could provide vital information about the direction of tsunami propagation and warn people about the approaching danger.

3.5. Quantum Computation

Current intense investigations in the field of applied quantum physics have produced a new step in quantum physics itself: instead of the usual stationary states (the states with definite energy) physicists study complex superpositions of stationary states. The "keyword" in contemporary physics has become "information" instead of energy.

The whole idea of quantum computation is based on the use of very sophisticated quantum states. Consider, for example, only two states of an atom. Let the ground state (the state with the minimum energy) be "g", and an excited state with higher energy be "e". According to quantum physics, the atom can be placed in a superposition of these two states. There are infinitely many such superpositions; the simplest of them can be written as (g+e). If you then try to measure the energy of the atom, you will get the energy corresponding to g or e with 50% probability. However, before such a destructive measurement, the atom exists partially in each of the two states, with two different energies, both at the same time — this is the basis for the miracle of quantum superposition!

Now let us consider two identical atoms: atom "1" and atom "2". The simplest superposition state for the two atoms is:

$$g_1 g_2 + g_1 e_2 + e_1 g_2 + e_1 e_2 \,.$$

Here index "1" refers to the first atom, and index "2" refers to the second atom. This state means that we have the two atoms simultaneously in both ground states, both atoms in the excited states, and both combinations of one atom in the ground state and the other atom in the excited state, everything and all at the same time! Thus, the two atoms hide $2^2 = 4$ stationary states. Now imagine 1000 atoms in the similar superposition state. They would hide $2^{1000} \approx 10^{301}$ stationary states.

This is often difficult to grasp and a visual aid to help appreciate the complexity of superposition states can be constructed using a thin sheet of clear plastic (e.g. an overhead projector slide). Draw some regularly spaced dark lines across the whole length of the slide and then cut it in half and overlay the two pieces. One can observe a whole range of new and different patterns by simply rotating one piece over the other. These Moiré patterns illustrate how two states (patterns) can be superimposed (overlayed) to produce a huge number of new states (patterns). Now imagine that all of these superpositions (patterns) exist simultaneously, and you get some idea of the amazing complexity of superposition states.

We can assign a definite number to every stationary state. For example, we can use a binary notation like the one used for conventional computers. In binary notation there are only two digits, "0" and "1". The number 2 can be written as (10), the number 3 as (11), the number 4 as (100), the number 5 as (101), and so on. If we assign the digit "0" to the ground state, then $(g_1 g_2)$ corresponds to zero; $(g_1 e_2)$ corresponds to one; $(e_1 g_2)$ corresponds to two, and $(e_1 e_2)$ corresponds to three. Now a superposition of the stationary states means a superposition of these numbers. With only 1000 atoms we can simultaneously represent a huge quantity (10^{301}) of numbers at the same time. Manipulating only 1000 atoms, one could manipulate 10^{301} numbers at the same time. This wonderful opportunity may allow scientists to solve some so-called "intractable" problems which cannot be solved with the most powerful conventional computers, including the factoring of very large numbers.

There are two very important challenges which are to be overcome on the way to practical quantum computation. The first one is the following: how to extract the hidden information from the atoms? For example, let 1000 atoms hide a huge amount of values of a function. How to extract the important properties of this function? Currently, a few brilliant quantum algorithms have been suggested which provide solutions to this challenge for some very important intractable problems. The second, more complicated challenge, is the following: how to preserve the fragile superposition states of many atoms and how to correct the inevitable errors? In spite of the great progress in the theory of quantum computation, a practical solution of this problem is far from being achieved.

3.6. Quantum Teleportation

Scientists who are working in the exciting area of quantum physics discuss and carry out experiments which remind many of mystic tales or science fiction stories rather than real scientific research. One class of these experiments is on quantum teleportation. The idea of teleportation is the following. Imagine that you have a nice drinking glass in one place and your friend has plenty of small pieces of glass at another place. These small pieces could be used to make the drinking glass, but it is a very complicated task. We can even assume that nobody knows how to do it. Next, you irradiate the drinking glass with electromagnetic pulses which destroy the drinking glass and you also make some simple measurements on the debris of your drinking glass. Then, you transfer to your friend the information derived from the measurements on the debris. Based on this information your friend irradiates the small pieces of glass with appropriate electromagnetic pulses. As a result, his pieces of glass are transformed into a drinking glass which is an exact copy of your drinking glass, while you retain the debris of the original drinking glass. It means you can interchange the positions of the drinking glass and the pieces of glass without any knowledge about how to make a drinking glass from the pieces of glass.

This is close to the idea of current experiments on quantum teleportation. The details of the experiments are rather complicated but the main idea is based on utilization of very sophisticated "entangled" states of two particles. The simplest entangled state for two atoms can be written as:

$$g_1 g_2 + e_1 e_2 \, .$$

This state means that either both atoms, "1" and "2", are in the ground state, or both atoms are in the excited state. There is a rigid correlation (or link) between the atoms: one of them cannot be in the ground state if the other one is in the excited state. It does not look very strange if the atoms are close to each other. However, imagine that these atoms are separated by a large distance! Two atoms separated by a large distance can still be tightly linked to each other. This connection is not the result of some special interaction but is an intrinsic property of quantum systems. If you make a destructive measurement transferring one atom of such a pair into the ground state, you can be certain that at the same moment of time the second atom, which is very far away, will also be immediately transferred into its ground state.

Such a mysterious link between entangled particles can be used to transfer information about the state of a third particle. Suppose, for example, that you wish to transfer the state of a third atom, "x", which is $cg_x + c'e_x$. Here, c and c' are arbitrary complex numbers with only one restriction: $|c|^2 + |c'|^2 = 1$, where $|c|^2$ and $|c'|^2$ are the probabilities for the atom to be found in the ground or excited states in a destructive measurement of energy. You do not need to know the state of your atom to transfer it! Suppose that you already have atom "1", and your friend has atom "2", and these two atoms are entangled. (You should create the entangled state for two initially close atoms and then allow the atoms to move far away from each other, all the time holding the entangled state, which is not a simple task.)

Next, you approach the third atom x with atom 1, so that they can interact to each other. You irradiate the x and 1 atoms with a few electromagnetic pulses and make a destructive measurement of their energy. As a result, the connection between the entangled atoms 1 and 2 will be broken, and the atom 2 will be in one of the following four states:

$$cg_2 + c'e_2, \quad \text{or} \quad c'g_2 + ce_2, \quad \text{or} \quad cg_2 - c'e_2, \quad \text{or} \quad c'g_2 + ce_2 .$$

These four possible states of atom 2 correspond to the four possible outcomes of your measurement on the atoms x and 1. (You can get $g_x g_1$ or $g_x e_1$, or $e_x g_1$, or $e_x e_1$.) Note: the key information about the numbers, c and c', is already transferred to atom 2! Next, you provide your friend, who is located near atom 2, with information about the results of your measurement. If the result of your measurement corresponds to the state $g_x g_1$, your friend knows that the state of the atom x was transferred to his atom 2. If the result of your measurement is different, say $e_x e_1$, your friend applies electromagnetic pulses which transfer the state $c' g_2 - c e_2$ into the state $c g_2 + c' e_2$. The required pulses are the same for any numbers c and c'. So, your friend can complete the transfer of the quantum state of x without any knowledge about the state itself (i.e. about the numbers c and c'.) So far, the experimental implementation of this brilliant idea has been achieved only for particles of light, i.e. photons. Instead of ground and excited states of atoms, scientists used vertical and horizontal polarizations of photons.

Chapter 4

SPACE

4.1. Normal Stars

The motions and properties of the planets, moons, stars, galaxies, and even the whole Universe are studied with the application of universal physical laws. Yet, scientists are very far from an understanding of space. Indirect data show that there are a great many "invisible objects", called "dark matter" which are difficult to study with conventional techniques. Probably the "simplest" objects to investigate are the stars, because they radiate visible light. The "normal stars", like our Sun, radiate energy due to fusion of hydrogen ^1_1H into helium ^4_2He in their cores. The sequence of fusion reactions that occur depends on the mass of the star. Scientists believe that for stars whose mass is close to or less than the mass of the Sun, three main reactions are responsible for the energy production. The first reaction is:

$$^1_1H + ^1_1H = ^2_1H + e^+ + \nu_e \,. \tag{4.1}$$

In this reaction, two protons overcome the repulsive electric interaction between them and fuse (due to the strong nuclear interaction) to form deuterium, ^2_1H, which consists of one proton and one neutron. This reaction occurs in the inner core of a star. As an example, in the core of the Sun the temperature is about 10 million kelvin (10^7 K). A typical speed for a proton is about a hundred thousand meters per second (10^5 m/s). As a result of collisions, some protons occasionally gain enough energy to approach each other closely enough to actually fuse together. Because the star consists mainly of individual protons (and electrons), it is estimated that about 10^{34} protons fuse every second in the Sun's core. As a result of this reaction, the kinetic energy of the particles increases. The positron, e^+, which is

43

the antiparticle of the electron, quickly "annihilates" with an electron producing two γ-ray photons which, in turn, will be scattered by other particles, increasing their kinetic energy.

The second reaction is:

$$^2_1H + ^1_1H = ^3_2He + \gamma. \tag{4.2}$$

Here a proton ^1_1H and deuteron ^2_1H fuse together producing the helium isotope, ^3_2He, and emitting a γ-ray photon. The third reaction,

$$^3_2He + ^3_2He = ^4_2He + ^1_1H + ^1_1H + \gamma, \tag{4.3}$$

produces the helium nucleus, ^4_2He, and two protons ^1_1H . Due to these three reactions, which are called the "proton–proton chain", the amount of helium nuclei, ^4_2He, increases with time.

Scientists believe that the proton–proton chain is responsible for the huge luminosity (light power emitted) of the Sun. At the same time, the fusion reactions support the structural stability of the normal star. Due to gravitational attraction, the star's particles tend to move toward the center of the star. Due to the fusion heating in the core of the star the pressure increases, forcing the particles outward from the core. These two processes reach an equilibrium where they are in balance with each other providing stability for the normal star. The luminosity of a normal star quickly increases with increasing mass. Correspondingly, the greater the mass of a normal star the shorter its lifetime. The mass of our Sun is approximately 2×10^{30} kg. Its luminosity is estimated as 4×10^{26} watt, and its lifetime is expected to be about 10 billion (10^{10}) years.

4.2.   Unusual Stars

Today many "unusual stars" are known, whose properties are very different from the properties of normal stars. Scientists believe that many of these unusual stars originated from normal stars, as a result of their natural evolution.

One type of well-known unusual star is the "White Dwarf." Their mass is about 60% of the Sun's mass, but their radius is about only

1% of the Sun's radius, i.e. close to the radius of the Earth! (The radius of the Sun is about 700 megameters (7×10^8 m), and the radius of the Earth is 6.4 megameters). Their surface temperature is greater than that of the Sun, but their luminosity typically is 100 times less than the Sun's luminosity. Scientists believe that the White Dwarfs originated from normal stars whose mass is close to or smaller than the Sun's mass.

The main idea of a White Dwarf's origin is the following. Eventually, most of the protons in the core of a normal star are converted into helium nuclei. The rate of the fusion reaction drops, and it cannot maintain the stability of the star's core anymore. So, the star's core shrinks in size due to the gravitational attraction. As a result, the kinetic energy of the helium nuclei (and, consequently, the temperature of the core) increases. When the temperature in the core reaches about 100 million kelvin (10^8 K) fusion of helium nuclei becomes possible. As a result of this fusion, helium nuclei are converted into carbon nuclei, increasing the kinetic energy of the particles and emitting γ-radiation . The star again becomes stable and continues to radiate light. At a temperature about 1 billion kelvin (10^9 K) carbon becomes a fuel, and so on. The fusion of nuclei less massive than iron can produce additional energy to heat the core of the star. The fusion of the iron and more massive nuclei requires energy and consumes some of the star's energy.

The change from hydrogen fuel in the core to helium fuel is accompanied by a change in the star's radius and color. The overall radius increases, and the star's color becomes red. The star becomes a "Red Giant". When the fuel in the star's core comes to the end, the star loses its outer layers ("envelope"), and the star's core contracts to a small size.

The next force which can fight the gravitational contraction is associated with the quantum properties of electrons. Electrons oppose the decrease in the star's volume not because of the electrostatic repulsion between them (the star is overall electrically neutral) but because of their half-integer spin: two identical particles having half-integer spin cannot occupy the same quantum state. This law is called the "Pauli Exclusion Principle" after W. Pauli. If the mass of the star is not very big, the pressure associated with the "electron gas" can withstand the gravitational attraction, producing a new but smaller equilibrium size. Such a star is observed as a White Dwarf. The White Dwarf does not generate additional energy. It just loses the initial energy collected in the process of the gravitational contraction and eventually becomes dim.

Unlike a normal star, the greater the mass of the White Dwarf, the smaller its radius. When the mass of a White Dwarf approaches 1.4 times the Sun's mass, its radius must approach zero! This means that if the mass of the star's core is greater than 1.4 of the Sun's mass the electron gas cannot withstand the gravitational attraction. Such massive stars normally explode, ejecting their envelopes and leaving the core, which contracts to a very small size. Such an explosion is observed as a "supernova" whose peak luminosity is comparable to the luminosity of a whole galaxy!

If the electron gas in the star's core cannot resist the gravitational attraction, the nuclei of the core break apart and the protons start to absorb electrons. In this reaction, protons p and electrons e^- are transformed into neutrons n and emitting electron neutrinos ν_e:

$$p + e^- \rightarrow n + \nu_e . \tag{4.4}$$

This process gives birth to the second well known type of unusual stars, the "neutron stars". In a neutron star the "neutron matter" resists the gravitational attraction, again because of the Pauli Exclusion Principle (the neutrons, like electrons, have spin 1/2). The radius of a typical neutron star ranges between 10 and 15 km. The greater the mass of the neutron star, the smaller its radius and so far it is unclear what the maximum possible mass is for a neutron star.

A typical neutron star rapidly rotates about its axis. Its period of rotation ranges from a few milliseconds to a few seconds. Scientists believe that "Pulsars", which radiate periodical electromagnetic pulses detected by radio-astronomers, are neutron stars. The time interval between neighboring electromagnetic pulses is believed to be equal to the period of the neutron star's rotation.

If the mass of the star's core after the supernova explosion exceeds the maximum mass possible for a neutron star, the "neutron matter" cannot resist the gravitational contraction. The gravitational contraction then gives the birth to the much discussed but very unusual and mysterious object called a "Black Hole". They are believed to be invisible point-like objects, which absorb any radiation or matter approaching them. A Black Hole is the source of a very strong gravitational field. Many stars (maybe even a whole galaxy) can orbit around a Black Hole like planets orbit a normal star.

4.3.   Space Flights

Space flights have become a routine part of our lives. Still they have growing importance for physics, technology and space exploration itself. One of the most important research laboratories is the International Space Station (ISS), which operates at a height of about 360 km above the ground. Scientists plan a lot of experiments in conditions of weightlessness on the board of the ISS.

The space shuttle Atlantis has already delivered the "Destiny Science Module" — the main science laboratory to the ISS. Destiny is approximately 8 m long and 4 m in diameter. ISS is expected to be completed in 2006, and its very important materials science laboratory will probably allow researchers to make crystals with very few defects.

A fleet of solar spacecrafts help scientists to study the "Sun's Envelope". While the nuclear reactions in the Sun's core have been explained, the outer part of the Sun remains a "puzzle" for scientists. The visible surface of the Sun is called the "photosphere". Above that surface is the "chromosphere", and then the "corona". The corona radiates the "solar wind" — the outward flow of electrically charged particles from the Sun. Every 11 years the Sun's surface becomes violent, producing firestorms which cause distortions in satellite and even ground based communications and power distribution systems. The temperature of the photosphere is about 6000 kelvin. Surprisingly, the temperature of the chromosphere is greater, about 10,000 kelvin, and the temperature of the corona is even higher than the temperature of the chromosphere: it reaches about 1,000,000 kelvin!

The solar spacecrafts are expected to help resolve the temperature and other puzzles of the Sun's corona. They detect visible, ultraviolet, and x-ray radiation from the Sun. As an example, the SOHO (Solar and Heliospheric Observatory) satellite orbits the Earth at a height of 1.5 billion meters, continuously monitoring the Sun's radiation.

A number of other spacecrafts study the mysteries of the planets, moons, asteroids, and comets orbiting our Sun. As an example, the data from the Galileo probe indicated the existence of liquid oceans under the surface of one of Jupiter's moons. Scientists suspect that these oceans could support some forms of life. The next flight to Jupiter which may verify these expectations is scheduled for 2008.

The Near Earth Asteroid Rendezvous (NEAR) spacecraft recently landed softly on the asteroid Eros at a distance of about 300 billion meters from the Earth. The photos obtained by NEAR showed that the surface of the asteroid is rocky. The NEAR's equipment could detect γ-radiation emitted by the asteroid and analyze the composition of the asteroid.

The space-based Chandra X-Ray Observatory, Multi-Mirror Mission and other x-ray satellites can detect the most violent events in the Universe: supernova explosions of stars, absorption of matter by Black Holes, and even collisions of stars. The cosmic x-rays cannot reach the Earth's surface and may only be studied from spacecraft. X-ray astronomy also indicates the existence of "magnetars" — neutron stars with the greatest known magnetic field, up to 100 gigatesla (10^{11} T). It also shows the existence of massive Black Holes at the centers of many galaxies.

Certainly, we have mentioned just a few of the space flight missions, and many other exciting programs are on the way.

Chapter 5

MECHANICS

5.1. Review

1. Classical Mechanics starts from the description of the motion of a "particle". A particle is a small, point-like, but macroscopic rigid solid entity. It contains many atoms but its physical size is negligible in a problem under consideration.

2. The "position" of such a particle can be described using a position vector $\vec{\mathbf{r}}$, which connects the origin of a Cartesian coordinate system with the particle. The vector $\vec{\mathbf{r}}$ is then described in terms of Cartesian coordinates of the particle

$$\vec{\mathbf{r}} = \vec{\mathbf{i}}x + \vec{\mathbf{j}}y + \vec{\mathbf{k}}z, \tag{5.1}$$

where $\vec{\mathbf{i}}$, $\vec{\mathbf{j}}$, and $\vec{\mathbf{k}}$ are "unit vectors" in the positive x, y, and z directions. The displacement of a particle $\vec{\boldsymbol{\Delta}}\mathbf{r}$ is given by

$$\vec{\boldsymbol{\Delta}}\mathbf{r} = \vec{\mathbf{r}}_f - \vec{\mathbf{r}}_i, \tag{5.2}$$

where $\vec{\mathbf{r}}_i$ and $\vec{\mathbf{r}}_f$ are the initial and final positions of the particle, respectively.

3. The "velocity" of a particle $\vec{\mathbf{v}}$ is the rate of change of its position

$$\vec{\mathbf{v}} = \frac{d\vec{\mathbf{r}}}{dt} = \vec{\mathbf{i}}v_x + \vec{\mathbf{j}}v_y + \vec{\mathbf{k}}v_z. \tag{5.3}$$

4. The "acceleration" of a particle $\vec{\mathbf{a}}$ is the rate of change of its velocity

$$\vec{\mathbf{a}} = \frac{d\vec{\mathbf{v}}}{dt} = \vec{\mathbf{i}}a_x + \vec{\mathbf{j}}a_y + \vec{\mathbf{k}}a_z. \tag{5.4}$$

5. The "average velocity" is defined as the average rate of change of the position

$$\vec{\mathbf{v}}_{av} = \frac{\vec{\mathbf{r}}_f - \vec{\mathbf{r}}_i}{\Delta t}, \tag{5.5}$$

where $\vec{\mathbf{r}}_i$ and $\vec{\mathbf{r}}_f$ are the initial and final positions in a problem under consideration, and Δt is the time interval of the motion.

6. The distance s traveled by a particle is the length of its trajectory, i.e. the path actually traveled by the particle.

7. The speed of a particle is the rate of increase of the distance traveled

$$\text{speed} = \frac{ds}{dt}. \tag{5.6}$$

The speed of the particle is equal to the magnitude of its velocity $\vec{\mathbf{v}}$.

8. The average speed of a particle is defined as the ratio $s/\Delta t$. In general, the average speed is not equal to the magnitude of the average velocity because of the vector nature of a finite displacement.

9. Any two particles affect each other through "interactions". If Particle 1 interacts with Particle 2, then Particle 1 exerts a "force" $\vec{\mathbf{F}}_{12}$ on Particle 2, and Particle 2 exerts a force $\vec{\mathbf{F}}_{21}$ on Particle 1.

10. The "Action–Reaction Principle" ("Newton's Third Law") states that $\vec{\mathbf{F}}_{12} = -\vec{\mathbf{F}}_{21}$.

11. The net force "$\vec{\mathbf{F}}$" acting on a particle is equal to the vector sum of all the forces acting on the particle.

12. The "Law of Inertia" ("Newton's First Law") states that if the

net force $\vec{\mathbf{F}}$ on a particle is zero, the particle retains its initial velocity $\vec{\mathbf{v}}_i$ relative to the remote stars (i.e. the particle retains its state of rest or uniform straight line motion relative to the remote stars).

13. The "Main Law of Mechanics" ("Newton's Second Law") states that the acceleration of a particle relative to the remote stars is proportional to the net force acting on the particle

$$\vec{\mathbf{a}} = \frac{\vec{\mathbf{F}}}{m}.$$ (5.7)

The quantity m is the "mass" of the particle. It describes the resistance a particle offers to acceleration (the greater the mass, the smaller the magnitude of the acceleration for the same magnitude of the net force).

14. The laws of mechanics are valid for all coordinate systems which move at constant velocity relative to the remote stars (such systems are called "inertial systems"). This statement is known as the "Relativity Principle".

15. The connection between the coordinates of two inertial systems S and S' are given by

$$\vec{\mathbf{r}} = \vec{\mathbf{r}}' + \vec{\mathbf{v}}_o t,$$ (5.8)

where $\vec{\mathbf{v}}_o$ is the velocity of the origin of the system S' relative to the origin of the system S. We suppose that the axes of S' are parallel to the axes of S and that time is measured from the moment when the origins of S and S' have the same position ($\vec{\mathbf{r}} = \vec{\mathbf{r}}'$ at $t = 0$). This transformation is known as the "Galilean Transformation".

16. The connection between the velocities in two inertial systems is given by

$$\vec{\mathbf{v}} = \vec{\mathbf{v}}' + \vec{\mathbf{v}}_o.$$ (5.9)

The acceleration of a particle is the same in both inertial systems ($\vec{\mathbf{a}} = \vec{\mathbf{a}}'$).

17. When studying mechanics we assume that particles do not have a net electric charge or a magnetic moment. In this case the only force between remote particles is the "gravitational force". However, when particles are in contact, they experience additional "contact forces".

18. "Newton's Law of Gravitation" states that the gravitational force exerted by particle 1 on particle 2 is given by

$$\vec{\mathbf{F}}_{12} = -G\frac{m_1 m_2}{r_{12}^2}\vec{\mathbf{n}}_{12}, \tag{5.10}$$

where G is the universal gravitational constant, r_{12} is the distance between the two particles, m_1 and m_2 are their masses and $\vec{\mathbf{n}}_{12}$ is the unit vector which points from particle 1 to particle 2. The negative sign shows that $\vec{\mathbf{F}}_{12}$ is opposite to $\vec{\mathbf{n}}_{12}$, i.e. particle 2 is attracted to particle 1.

19. A large solid rigid object can be considered as a system of many particles connected to each other. The position of such an object is described by the position $\vec{\mathbf{r}}$ of its "center of mass".

$$\vec{\mathbf{r}} = \frac{\sum m_n \vec{\mathbf{r}}_n}{m} \approx \frac{\int \vec{\mathbf{r}}'\, dm'}{m}, \tag{5.11}$$

where m is the mass of the object.

20. The center of mass of any system of particles (in particular, the center of mass of a solid object) moves like a single particle according to the main law of mechanics

$$\vec{\mathbf{F}} = m\vec{\mathbf{a}}, \tag{5.12}$$

where $\vec{\mathbf{F}}$ is the total net force on the system of particles, m is its total mass, and $\vec{\mathbf{a}}$ is the acceleration of the center of mass.

21. The gravitational force between two uniform spherical objects is given by the Newton's Law of Gravitation (5.10), where r_{12} becomes the distance between the centers of the spherical objects, and $\vec{\mathbf{n}}_{12}$ is the unit vector, which points from the center of object 1 to the center of object 2.

22. For "terrestrial phenomena" the characteristic time is often very much smaller than one day. In this case we can consider a coordinate system connected to the ground as an inertial system. For terrestrial phenomena, the magnitude of the gravitational force produced by the Earth on an object of mass m can be approximated as

$$F_g = mg \, , \tag{5.13}$$

where g is the gravitational acceleration.

23. If a flat surface of one object is in contact with a flat surface of another object, the contact force exerted by object 1 on object 2 can be represented as a sum of two forces: the normal force $\vec{\mathbf{F}}_n$ (the force of repulsion, which is perpendicular to the surfaces), and a tangential friction force.

24. If object 2 is sliding relative to object 1 the tangential force on object 2 is called the "kinetic friction force" $\vec{\mathbf{F}}_k$. This force is opposite to the direction of the relative velocity ($\vec{\mathbf{v}}_2 - \vec{\mathbf{v}}_1$). The magnitude of this force is related to the magnitude of the normal force by the approximate equation

$$F_k = \mu_k F_n \, , \tag{5.14}$$

where μ_k is called "the coefficient of kinetic friction".

25. If object 2 is not sliding relative to object 1 the tangential force, which prevents sliding, is called the "static friction force" $\vec{\mathbf{F}}_s$. The magnitude of F_s cannot be greater than the product $\mu_s F_n$ ($F_s \leq \mu_s F_n$), where μ_s is called "the coefficient of static friction".

26. Let an object be attached to a rope (or string or wire). If the rope is taut, the force exerted by the rope on the object points in the direction of the rope. This force is often called the "tension force".

27. Let an object be attached to a spring, which can be compressed or stretched, for example, in the x direction. The force exerted by

the spring on the object is given by

$$\vec{F} = -\vec{i}k_s(x - x_0),$$
(5.15)

where k_s is the spring (or force) constant of the spring, x is the position of the end of the spring, and x_0 is its equilibrium position. (We suppose that the other end of the spring is fixed at $x < x_0$.) This force is called the "spring force" or the "elastic force". If $x > x_0$ the spring force points in the direction of the spring, if $x < x_0$ it points in the opposite direction.

28. If the net force $\vec{F}$ on a particle is constant, the position of the particle is given by

$$\vec{r} = \vec{r}_i + \vec{v}_i t + \frac{\vec{a}t^2}{2}, \quad \vec{a} = \frac{\vec{F}}{m},$$
(5.16)

where $\vec{r}_i$ and $\vec{v}_i$ are the position and velocity of the particle at time $t = 0$. The velocity $\vec{v}$ of the particle is given by

$$\vec{v} = \vec{v_i} + \vec{a}t.$$
(5.17)

Two other expressions are useful

$$\vec{r}_f = \frac{1}{2}(\vec{v}_i + \vec{v}_f)t_f,$$
(5.18)

$$v_{\alpha f}^2 = v_{\alpha i}^2 + 2a_\alpha(\alpha_f - \alpha_i),$$

where α is x, or y, or z, and the index "f" means final.

29. The "momentum" of a particle $\vec{p}$ is defined as the product of its mass and velocity

$$\vec{p} = m\vec{v}.$$
(5.19)

The rate of change of the momentum is equal to the net force on the particle

$$\frac{d\vec{p}}{dt} = \vec{F}.$$
(5.20)

The same equation is valid for a system of particles, where

$$\vec{\mathbf{p}} = \sum m_n \vec{\mathbf{v}}_n \tag{5.21}$$

is the total momentum of the system, and $\vec{\mathbf{F}}$ is the total net force on the system.

30. If the net force on a system of particles equals zero, the momentum of the system does not change and so is conserved (the "Law of Conservation of Momentum"). If any component of the net force equals zero, the same component of the momentum is conserved.

31. The impulse delivered by a force $\vec{\mathbf{F}}_k$ to a particle during the time interval $t_i < t < t_f$ is defined as

$$\vec{\mathbf{I}}_k = \int_{t_i}^{t_f} \vec{\mathbf{F}}_k \, dt \,. \tag{5.22}$$

The impulse delivered by a net force equals the change in the particle's momentum ("The Impulse–Momentum Theorem")

$$\vec{\mathbf{I}} = \vec{\mathbf{p}}_f - \vec{\mathbf{p}}_i \,. \tag{5.23}$$

32. The kinetic energy of a particle is defined as

$$K = \frac{mv^2}{2} \,. \tag{5.24}$$

The rate of change of the kinetic energy is given by

$$\frac{dK}{dt} = \vec{\mathbf{F}} \cdot \vec{\mathbf{v}} \,. \tag{5.25}$$

The vector dot product $\vec{\mathbf{F}} \cdot \vec{\mathbf{v}}$ is called the "power" of the force $\vec{\mathbf{F}}$.

33. The work done by a force $\vec{\mathbf{F}}_k$ on a particle moving from the point P_i to the point P_f is defined as

$$W = \int_{P_i}^{P_f} \vec{\mathbf{F}}_k \cdot d\vec{\mathbf{r}} \,. \tag{5.26}$$

The work done by the net force is equal to the change in the particle's kinetic energy ("Work–Kinetic Energy Theorem")

$$W = K_f - K_i \, . \tag{5.27}$$

34. The work done by the gravitational force on a particle of mass m near the surface of the Earth (terrestrial phenomenon) is given by

$$W_g = mg(y_i - y_f) \, , \tag{5.28}$$

where y_i and y_f are the initial and final coordinates of the particle, assuming that the y-axis points vertically upwards. The function

$$U_g(y) = mgy + \text{constant} \tag{5.29}$$

is called the gravitational potential energy.

35. The kinetic energy of a system of particles is defined as the sum of the kinetic energies of all the particles

$$K = \sum K_n \, . \tag{5.30}$$

The gravitational potential energy of a system of particles is the same as for a single particle, $U(y) = mgy + \text{constant}$, where m becomes the total mass of the system, and y becomes the y-coordinate of its center of mass.

36. The work done by a spring force on an object, which is in contact with the spring's free end and is moving along the spring axis (x-axis), is given by

$$W_s = \frac{1}{2}k_s(x_i^2 - x_f^2), \tag{5.31}$$

where x_i and x_f are initial and final positions of the spring end. The function

$$U_s(x) = \frac{1}{2}k_s x^2 + \text{constant} \tag{5.32}$$

is called the spring (or elastic) potential energy.

37. The mechanical energy E of a system of particles (or a simple solid object) is defined as the sum of its kinetic and potential energies

$$E = K + U .   \tag{5.33}$$

If only gravitational and spring forces do work on the system of particles, then the mechanical energy of the system is conserved (the "Law of Conservation of Mechanical Energy"). The gravitational and spring forces are called "conservative forces". The work done on a particle by conservative forces depends only on the initial and final positions of the particle.

38. Let a particle move around a circle of radius r in a circular motion. The acceleration of the particle has two components: the centripetal component, $\vec{a}_c$, which points towards the center of the circle and the tangential component, $\vec{a}_t$, which is tangential to the circle. The magnitude of the centripetal component is given by

$$a_c = \frac{v^2}{r} ,   \tag{5.34}$$

where v is the speed of the particle and the tangential component is zero if v is constant.

39. Let a particle revolve about the z-axis, in the x–y plane. The angular position of the particle is given by

$$\vec{\theta} = \vec{\mathbf{k}}\theta ,   \tag{5.35}$$

where θ is the angle between the position vector $\vec{\mathbf{r}}$ of the particle and the x-axis. The angular displacement $\vec{\Delta}\theta$ can be described by the vector

$$\vec{\Delta}\theta = \vec{\theta}_f - \vec{\theta}_i = \vec{\mathbf{k}}(\theta_f - \theta_i) .   \tag{5.36}$$

The angular velocity $\vec{\omega}$ is defined as

$$\vec{\omega} = \frac{d\vec{\theta}}{dt} ,   \tag{5.37}$$

and its magnitude is the angular speed ω.

The angular acceleration $\vec{\alpha}$ is

$$\vec{\alpha} = \frac{d\vec{\omega}}{dt} \, . \tag{5.38}$$

40. The connections between v, ω, a_t and α are given by

$$v = r\omega, \quad a_t = r\alpha \, . \tag{5.39}$$

41. Let a solid object rotate about a fixed axis (z-axis). The position of the rotating object can be described by the angle θ between the x-axis and the position vector for any point in the object revolving in the x–y plane (x–y plane "cuts" the object). Angular position, displacement, velocity and acceleration for the rotating object are defined in the same way as for a particle.

42. The torque $\vec{\tau}_k$ produced by the force $\vec{\mathbf{F}}_k$ is defined as the vector cross product:

$$\vec{\tau}_k = \vec{\mathbf{r}}_k \times \vec{\mathbf{F}}_k \, , \tag{5.40}$$

where $\vec{\mathbf{r}}_k$ is the position vector of the point of application of the force.

43. The moment of inertia I of a particle revolving about the z-axis is defined as

$$I = m(x^2 + y^2) \, . \tag{5.41}$$

The moment of inertia of a solid object rotating about the z-axis is

$$I = \int (x^2 + y^2) dm \, . \tag{5.42}$$

44. The "Law of Rotation" is similar to the Main Law of Mechanics and is also know as "Newton's Second Law for Rotation"

$$\vec{\alpha} = \frac{\vec{\tau}}{I} \, , \tag{5.43}$$

where $\vec{\tau}$ is the net torque (vector sum of all torques) acting on the object.

45. The angular momentum of a particle is defined as the cross product

$$\vec{\mathbf{L}} = \vec{\mathbf{r}} \times \vec{\mathbf{p}}\,. \tag{5.44}$$

For a solid object rotating about a fixed axis, the angular momentum is

$$\vec{\mathbf{L}} = I\vec{\omega}\,. \tag{5.45}$$

46. The rate of change of a particle's angular momentum is equal to the net torque on the particle

$$\frac{d\vec{\mathbf{L}}}{dt} = \vec{\tau}\,. \tag{5.46}$$

The same equation is valid for a system of particles, and in particular, for a solid object.

47. If the net torque on a system of particles equals zero, the angular momentum of the system does not change (the "Law of Conservation of Angular Momentum").

48. The rotational kinetic energy of a revolving particle can be written as

$$K = \frac{I\omega^2}{2}\,. \tag{5.47}$$

The same equation is valid for a rotating solid object.

49. The conditions for static equilibrium of an object are the following:

$$\vec{\tau} = 0\,, \quad \vec{\mathbf{F}} = 0\,, \tag{5.48}$$

assuming $\vec{\mathbf{v}}_i = 0$ and $\vec{\omega}_i = 0$.

50. To describe celestial phenomena in the solar system it is convenient to use a coordinate system connected to the center of the Sun. It is a very good approximation to take it as an inertial system. To describe the motion of a satellite about the Earth it is convenient to use a coordinate system connected to the Earth's center. For the

satellite's motion it can be considered an inertial system.

51. Planets orbit the Sun moving along elliptical paths ("Kepler's First Law"). The ratio of the squared period T^2 to the cubed semi-major axis R_1^3 is the same for all planets ("Kepler's Third Law").

52. If the net force on an object is the spring force $\vec{F} = -\vec{i}k_s(x - x_0)$, the motion of the object (rigorously, the motion of the end of the spring which is attached to the object) is given by

$$x = x_0 + A\cos(\omega t + \phi).\qquad(5.49)$$

This equation describes the harmonic oscillation of an object with amplitude A, angular frequency ω and phase constant ϕ. The frequency of the oscillation is given by

$$f = \frac{\omega}{2\pi},\qquad(5.50)$$

and the period T, is

$$T = \frac{1}{f} = \frac{2\pi}{\omega}.\qquad(5.51)$$

53. A disturbance traveling through a substance without permanent displacements of the particles of the substance is called a traveling wave. If the disturbance can be described by a harmonic function (sine or cosine) the traveling disturbance is called a harmonic traveling wave.

54. The simplest example of a harmonic traveling wave is a wave on a long taught string. If one of the ends of the string starts to move in a simple harmonic oscillation, e.g. along the y-axis, the oscillation "travels" along the string in the x-direction, if x-axis points along the string (see Fig. 7).

55. The y-coordinate of a point on a string in a harmonic traveling wave is given by

$$y(x, t) = A\sin(kx - \omega t + \phi), \quad 0 \le x \le \frac{\omega t}{k},\qquad(5.52)$$

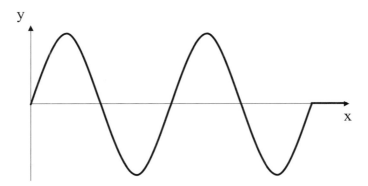

Fig. 7.   A harmonic traveling wave.

if the wave travels in the positive x-direction or

$$y(x,t) = A\sin(kx + \omega t + \phi), \quad -\frac{\omega t}{k} \le x \le 0, \qquad (5.53)$$

if the wave travels in the negative x-direction. (We assume that the wave is produced at the origin beginning at time $t = 0$.)

56. The quantity k is called the wave number, ω is the angular frequency, ϕ is the phase constant, and A is the amplitude of the wave. The angle $(kx \mp \omega t + \phi)$ is called the phase of the wave. These parameters are common for all harmonic traveling waves. Points where the displacement from the equilibrium position is the largest are called the crests of the wave.

57. The frequency of a harmonic wave is defined as

$$f = \frac{\omega}{2\pi}. \qquad (5.54)$$

58. The period of a harmonic wave

$$T = \frac{1}{f} = \frac{2\pi}{\omega}. \qquad (5.55)$$

59. The wavelength λ of a harmonic wave is the distance between two neighboring points of the same phase

$$\lambda = \frac{2\pi}{k}. \qquad (5.56)$$

60. The speed of a harmonic wave

$$v = \frac{\omega}{k} = \frac{\lambda}{T} = \lambda f . \qquad (5.57)$$

All these relations are common for all harmonic traveling waves.

61. The speed of a traveling wave in a string is given by

$$v = \sqrt{\frac{F_T}{\mu}} , \qquad (5.58)$$

where F_T is the tension force in the string and $\mu = m/L$ is the linear mass density of the string (of mass m and length L).

62. The power transmitted by a harmonic traveling wave in a string is given by

$$P = \frac{1}{2}\mu v (A\omega)^2 . \qquad (5.59)$$

63. The intensity of a sound wave, I_S is the power transmitted by the sound wave P_S per unit area

$$I_S = \frac{dP_S}{dA} , \qquad (5.60)$$

where a surface of area dA is assumed to be perpendicular to the direction of propagation of the sound wave.

64. The level of a sound wave β is defined as

$$\beta = 10 \log \frac{I_S}{I_{SO}} , \qquad (5.61)$$

where $I_{SO} = 10^{-12}$ W/m^2 is conventionally called the "threshold of hearing."

65. When a source of a sound wave and (or) an observer who detects a sound wave, move relative to the air, the frequency of the detected wave is given by (the "Doppler Effect")

$$f' = \frac{v \pm v_o}{v \mp v_s} f , \qquad (5.62)$$

where f is the frequency of the stationary source detected by a stationary observer, v is the speed of sound, v_o is the speed of the observer, v_s is the speed of the source, the upper signs (increase of the frequency) correspond to the source moving toward the observer and the observer moving toward the source, the lower signs correspond to the opposite cases.

66. A harmonic standing wave in a string is described by the equation for the y-coordinate of a point in the string

$$y = A \sin (kx + \phi_1) \cos (\omega t + \phi_2).$$ (5.63)

67. The amplitude of a harmonic standing wave $A \sin (kx + \phi_1)$ is a harmonic function of x. The points which do not oscillate are called "nodes," the points which oscillate with the maximum amplitude A (amplitude of the standing wave) are called "antinodes." All points between two neighboring nodes oscillate with the same phase $(\omega t + \phi_2)$ or $(\omega t + \phi_2 + \pi)$.

68. The wavelength of the harmonic standing wave, $\lambda = 2\pi/k$, is the distance between two neighboring points of the same amplitude and phase. The standing wave does not have a speed because the oscillations "fill" the whole string and the nodes and antinodes do not move along the string. The period of the wave $T = 2\pi/\omega$, and the frequency $f = 1/T = \omega/2\pi$. The possible frequencies of the harmonic standing waves take on discrete values f_n. The minimum frequency is called the fundamental frequency, and the corresponding standing wave is called the fundamental mode. All frequencies f_n are called natural frequencies. All these properties and terms are common for all harmonic standing waves.

69. For a string fixed at both ends the fundamental frequency is

$$f_1 = \frac{v}{2L},$$ (5.64)

where L is the length of the string, and v is the speed of the harmonic traveling wave in the string. All natural frequencies are given by

$$f_n = nf_1, \quad \text{where} \quad n = 1, 2, 3, \ldots$$ (5.65)

The corresponding natural wavelengths are

$$\lambda_n = \frac{v}{f_n} = \frac{2L}{n}, \quad \text{where} \quad n = 1, 2, 3, \ldots \tag{5.66}$$

70. For a string fixed at one end the natural frequencies are given by

$$f_n = nf_1, \quad n = 1, 3, 5, \ldots, \tag{5.67}$$

where the fundamental frequency is

$$f_1 = \frac{v}{4L}. \tag{5.68}$$

The corresponding natural wavelengths are

$$\lambda_n = \frac{v}{f_n} = \frac{4L}{n}, \quad n = 1, 3, 5, \ldots \tag{5.69}$$

5.2. Problems

1. The troposphere (where the weather takes place) is the lower part of the Earth's atmosphere, extending for approximately 10 km above the ground. Estimate the relative change in the gravitational acceleration in the troposphere (in percent).

2. The presence of bacteria at a height of about 40 km above the ground was discovered in 2001. Scientists suppose the bacteria arrive from outer space because nothing can transport them from the ground to such a great height. Estimate the gravitational force on a bacteria at a height of 40 km, assuming its mass to be of the order of 1 fg.

3. The Voyager-2 probe was launched in 1977. It passed near Jupiter, Saturn, Uranus, and Neptune. Now it moves out of the solar system.

a) The speed of the Voyager-2 spacecraft is about 38,000 mph. Find its speed in SI units.

b) Assuming a constant speed, estimate the distance traveled by Voyager-2 in one year.

4. The Cosmos-1 project, privately funded by the Planetary Society, is designed to use the pressure of the Sun's radiation to move a solar sail spacecraft like an ordinary sail uses the pressure of the wind.

a) The area of the Cosmos-1 sail is about 720 square yards, and the weight of the spacecraft is about 88 pounds. Find the area and mass in SI units.

b) What must be the magnitude of the force exerted by the solar light on the sail to balance the gravitational force on the spacecraft exerted by the Earth at a distance of 100 Mm from the center of the Earth?

c) Let the solar sail spacecraft move away from the Earth and the Sun. At what distance from the center of the Earth will the gravitational force on the spacecraft exerted by the Earth be equal to the gravitational force exerted by the Sun? (Assume the Sun, the Earth and the spacecraft are on the same line, in this order.)

5. The interstellar solar sail probe is expected to be launched by 2010. It is estimated that it can reach a speed of 380,000 mph, 10 times greater than Voyager-2 (see problem 3).

a) Estimate the distance the interstellar probe can cover in one year.

b) Estimate the momentum and kinetic energy of the interstellar probe, assuming its mass is 40 kg.

6. The Galileo mission, which started in 1989, consisted of a 2400 kg orbiter and a 300 kg probe. The orbiter became the first spacecraft to orbit Jupiter. The Galileo probe traveled first toward the Sun, encountered Venus and then the Earth twice, so as to gain enough speed to reach Jupiter. Consider an encounter with the Earth at a distance of about 1 Mm from the ground.

a) Find the gravitational force (magnitude) exerted by the Earth on the spacecraft.

b) Estimate the magnitude of the acceleration of the spacecraft.

7. A spacecraft, Genesis, is to collect atoms from the solar wind and deliver them to the Earth. The distance to be covered in this round

trip mission is about 20 million miles and the expected time for the trip is about 3 years. Estimate the average speed of the Genesis craft in SI units.

8. The Genesis spacecraft is going to collect solar atoms in a region which is at a distance about 92 million miles from the center of the Sun and 1 million miles from the center of the Earth.

Estimate the ratio of the gravitational force exerted by the Sun on the spacecraft to the gravitational force exerted by the Earth in that region. (Consider only the magnitudes of the forces.)

9. It is expected that atoms of all the naturally occurring elements will be collected by the Genesis spacecraft. Assume that a gold atom in the solar wind moving at 10^6 mph relative to the spacecraft is captured by a Genesis collector.

a) Find the magnitude of the impulse delivered by the atom to the spacecraft.

b) Is the total momentum of the spacecraft and the atom conserved in the collision?

c) Find the kinetic energy of the gold atom relative to the spacecraft before the impact. Is the total kinetic energy of the spacecraft and the atom conserved in the impact?

10. Assume that an iron atom in the solar wind travels at 10^6 mph relative to the center of the Earth.

a) If it moves along a line which passes through the center of the Earth, find the magnitude of its angular momentum relative to the center of the Earth.

b) Solve the same problem if it moves along a line which is tangential to the Earth's surface.

11. The Cassini spacecraft launched in 1997 is expected to arrive at Saturn in 2004. Its orbiter will orbit Saturn to study its rings and moons. Estimate the gravitational force between the Sun and the 2000 kg orbiter traveling around Saturn. Take the distance between Saturn and the Sun to be 9.5 astronomical units (AU).

12. Cassini is to deliver a probe called Huygens to Saturn's moon Titan. Estimate the gravitational acceleration near Titan's surface. Take the radius of Titan to be 2.6 Mm and its mass to be 1.35×10^{23} kg.

13. The 12,600 pound Cassini spacecraft passed Jupiter at a distance of 6 million miles from its center in December 2000. Find the gravitational force between the spacecraft and Jupiter at this distance.

14. Images taken by Cassini confirmed the existence of long-lived Jovian storms, which can last many years and probably several centuries!

a) Assuming 300 mph winds in the Jovian atmosphere, find the momentum (magnitude) and kinetic energy of an ammonia molecule (NH_3) moving with the speed of the wind.

b) Estimate the angular momentum of the molecule (magnitude) relative to the center of Jupiter. Take the radius of Jupiter to be 7.18×10^7 m.

15. A comet nucleus is a small celestial body consisting of ice and dust. Assuming that a comet is a pure ice ball of 3 km diameter, estimate its mass. Take the density of ice to be 917 kg/m^3.

16. As a comet approaches the Sun it releases gas and dust under the action of the Sun's radiation and forms a "coma". Assuming that the mass of the coma is 10% of the total mass estimated in problem 15, and it has a spherical form 100 megameters in diameter, estimate the average density of the coma.

17. Typically a comet orbits the Sun in a highly elliptical orbit. "Long-period comets" have periods of more than 200 years. Some of them have periods of up to 10^7 years! Assume that a comet has a perihelion distance (the closest distance to the Sun) of 3 astronomical units (AU) and the aphelion distance (the maximum distance from the Sun) 10^5 AU.

a) As the first approximation, consider its motion to be a straight line oscillation and estimate the average speed of the comet.

b) Find the ratio of gravitational forces (magnitudes) between the comet and the Sun at aphelion and perihelion.

18. In July 1994 the comet "Shoemaker–Levy" collided with Jupiter. It is estimated that long before the collision the comet nucleus split into about 20 fragments. Assume that a spherical pure ice fragment 1 km in diameter crashed into Jupiter at 60 km/s. Estimate the kinetic energy and the magnitude of the fragment's momentum just before the collision. Estimate the magnitude of impulse delivered by the fragment to Jupiter. Take the density of ice to be 917 kg/m^3.

19. It is estimated that the nucleus of the Shoemaker–Levy comet split apart under the action of "tidal forces" (i.e. the nonuniformity of the gravitational force across a body) at a distance of about 20 Mm from Jupiter. Estimate the difference between the gravitational force magnitudes produced by Jupiter on two particles of the same mass m on the surface of the comet but assuming that one particle is 5 km closer to Jupiter's center than the other one.

20. About 10 times per year the Earth passes through swarms of cosmic dust particles, causing "meteor showers". A piece of such cosmic dust is called a "meteoroid". The typical size of a meteoroid is 1 cm, and its mass is about 1 g. Meteoroids burn up in the Earth's atmosphere producing meteors ("shooting stars"). A typical meteoroid begins to shine at a height of about 90 km above the ground.

a) Estimate the density of a typical meteoroid.

b) Estimate the acceleration of a meteoroid at a height of 90 km.

c) Assume a meteoroid "collides" with the Earth's atmosphere at a speed of 50 km/s. Estimate the magnitude of the momentum and the kinetic energy of such a typical meteoroid.

21. The solar-powered unmanned aircraft Helios is expected to reach an altitude of about 10^5 ft. The current altitude record for a propeller-drive plane is about 8×10^4 ft. Find the difference between these two altitudes in meters.

22. The 62,000 solar cells on the Helios aircraft generate about 40 kW of power to drive its 14 propellers. How much energy is needed to drive one propeller for an hour?

23. Assume the Helios aircraft remains approximately above the same point on equator at a height of 10^5 ft.

a) Find the speed of the aircraft relative to the center of the Earth.

b) Find the magnitude of acceleration of the aircraft relative to the center of the Earth.

c) Find the ratio of the gravitational force exerted on the aircraft by the Earth to the total force exerted on it by the air (magnitudes). (Note that the net force on Helios is sum of the gravitational force and the total force exerted by the air.)

d) Find the ratio of the aircraft's angular momentum to its momentum (magnitudes) in meters.

24. Assume the Helios aircraft travels at 150 mph relative to the ground. Find the ratio of its kinetic energy to the magnitude of its momentum (in m/s) relative to the ground.

25. In experiments with a high energy beam of lead ions colliding with stationary lead ions, the matter density was estimated to be 20 times that of normal nuclear (proton) matter. Estimate the density achieved in such experiments (in kg/m^3). Use 1.5 femtometers as the radius of a proton.

26. The normal Atlantic hurricane season lasts from June until December. Typically it brings two or three major hurricanes with wind speeds greater than 110 mph. What is the kinetic energy (in joules) of 1 m^3 of air, if it moves at 120 mph? Take the density of air to be 1.3 kg/m^3.

27. In problem 26 what is the magnitude of the momentum of 1 m^3 of air (in Ns)?

28. A major hurricane "supports" winds greater than 110 mph. Find the kinetic energy in joules of a nitrogen molecule moving with the

speed of the wind, 110 mph.

29. The NEAR spacecraft softly landed on the asteroid Eros. The landing speed of the 1,100 pound NEAR did not exceed 1.8 m/s. Assuming this value for the speed estimate:

a) The momentum (magnitude) and kinetic energy of the spacecraft.

b) The impulse (magnitude) delivered by Eros' surface on the spacecraft.

30. Assume that the international space station ISS "Alpha" module orbits the Earth at an altitude of 230 miles above the ground. Assuming a circular orbit find the speed, angular speed, acceleration (magnitude), and orbital period of Alpha.

31. If the distance between ISS Alpha and the center of the Earth increases 9 times, how would it affect the speed, angular speed, acceleration, and orbital period of Alpha?

32. If the distance between ISS Alpha and the center of the Earth increase 9 times, how would it affect the momentum, angular momentum, and kinetic energy of Alpha?

33. The main science laboratory and command center of Alpha is called "Destiny". It is 8.4 m long, and has a diameter of 4.2 m. If the weight of Destiny is 30,000 pounds, what is its average density (in kg/m^3)?

34. Taking data from problems 30 and 33 find:

a) The gravitational force (magnitude) exerted by the Earth on Destiny.

b) The momentum, angular momentum (magnitudes) and kinetic energy of Destiny.

35. Estimate the ratio of the gravitational forces exerted by the Earth on Alpha to the average force exerted by the Sun on Alpha (magnitudes). Take the altitude of Alpha to be 230 miles.

36. The Boeing Delta 4 launch vehicle is expected to launch payloads of mass about 13,000 kg. Assume a Delta 4 delivers a 13 Mg payload to a geosynchronous orbit (the period of the geosynchronous orbit is one day). Find the final momentum, angular momentum and kinetic energy of the payload.

37. In 2000 astronomers discovered a small planet ("Plutino") 2000EB173 between Neptune and Pluto. Plutino, which is about 400 miles in diameter, orbits the Sun with a period estimated to be 240 years. Estimate the semimajor axis of Plutino's orbit (in astronomical units).

38. The most distant galaxies scientists can now see are about 10 billion lightyears (10^{10} ly) away. Estimate the volume of the visible universe (in m^3).

39. A typical "large" asteroid with a size of 100 km and density about 2.5 g/cm^3, spins at 5 revolutions per day. Assuming that the asteroid has an approximately spherical shape, estimate (in a coordinate system connected to the asteroid's center):

    a) the angular speed of the asteroid (in rad/s);

    b) the speed of a point near the asteroid's equator;

    c) the magnitude of the angular momentum (the moment of inertia for a uniform sphere is 2 $mr^2/5$);

    d) the rotational kinetic energy.

40. Solve problem 39:

    a) for a very large asteroid of size 800 km, which spins at 2 revolutions per day;

    b) for a very small asteroid of size 50 m, which spins at 500 revolutions per day.

41. The asteroid Castalia has a so-called Earth crossing orbit. In 1989 it was detected by radar at a distance of about 11 lunar distances from the Earth. Castalia's dimensions have been found to be about $1.8 \times 0.8 \times 0.8$ km, with a density of 2.1 g/cm^3 and a rotational period of 4 hours. Based on these data estimate:

   a) Castalia's mass (assume it has a cylindrical shape);
   b) the gravitational force between Castalia and the Earth at 11
lunar distances from the Earth;
   c) the angular speed of Castalia's rotation.

42. On its way to Jupiter the Galileo probe passed through the main
asteroid belt (between the orbits of Mars and Jupiter). Unexpect-
edly, Galileo detected an asteroid with a satellite, "Dactyl", which
orbits the asteroid "Ida". The dimensions of Ida are approximately
$60 \times 20 \times 20$ km and its density is 2.5 g/cm^3. The size of Dactyl is
about 1.4 km. Assuming the same density for Ida and Dactyl find
the ratio of the gravitational forces exerted by the Sun on the two
celestial bodies. Assume a cylindrical shape for Ida and a spherical
shape for Dactyl.

43. The dark space in our Galaxy (the "Milky Way") is associated
with "clouds" of interstellar dust. A typical particle of dust, which
blocks visible light, is an elongated grain with dimensions about
$0.4 \times 0.2 \times 0.2$ μm. It consists of a rocky core and an ice mantle
and has a mass of about 10^{-14} g. Find the average density of a grain
assuming it has a cylindrical shape.

44. It is estimated that on average every 10^6 m^3 of interstellar space
contains one dust particle. Using data from problem 43, estimate:
   a) The average distance between nearest neighbor dust grains.
   b) What part of interstellar space is occupied by dust particles?
   c) The average density of matter in interstellar space.
   d) The average gravitational force between two nearest neighbor
dust particles.

45. The cantilever used in magnetic resonance force microscopy is
a tiny "beam", which is fixed at one end and free to vibrate at the
other end. The free end of the cantilever oscillates approximately
in a simple harmonic motion like a small block attached to a mass-
less spring. It can be described by an effective mass and an effective
spring constant.
   Consider an ultra-thin silicon cantilever (thickness about 60 nm,

length about 200 μm). If the frequency of the cantilever vibrations is 1.7 kHz, and the effective spring constant is 6.5×10^{-6} N/m, find the angular frequency, the period and the effective mass of the cantilever.

46. Let the cantilever tip in problem 45 oscillate along the x-axis near the origin. At time $t = 0$ let the cantilever tip be at the end-point of its motion at $x = 40$ nm.

a) Find the position, velocity and acceleration of the cantilever tip at time $t = 1$ ms.

b) When does the cantilever tip pass through the origin for the first time, and for the second time?

c) When does the cantilever tip pass through the point $x = 20$ nm for the first time, and for the second time?

47. Let the cantilever tip in problem 45 oscillate with an amplitude of 30 nm.

a) What is the maximum speed of the cantilever tip?

b) What is the speed of the tip at a distance of 10 nm from the equilibrium position?

c) What are the maximum magnitudes of the acceleration and the net effective force on the cantilever tip?

d) What are the magnitudes of the acceleration and the net force at a distance of 10 nm from the equilibrium position?

48. Assume that the magnetic moment of a single proton produces a magnetic force of magnitude 6×10^{-19} N on a small magnet attached to a cantilever tip. If the force acts in the direction of the maximum possible displacement of the tip, estimate the magnitude of its equilibrium displacement. Take data from problem 45.

49. It is estimated that a maglev (a train, which is levitated due to the action of magnetic forces) could achieve a speed 500 km/h. If the acceleration of the train is 1/5 of the gravitational acceleration, how long does it take to achieve this speed starting from rest?

50. In problem 49, find the distance traveled by the train before it achieves a speed of 500 km/h.

51. In problem 49, find the magnetic force (magnitude) needed to levitate a 55,000 kg train car.

52. In problem 49, assume that the power dissipated by air resistance at a speed of 500 km/h is about 10 MW. Estimate the force of air resistance on the train.

53. In problem 49, find the kinetic energy and momentum (magnitude) of a 55,000 kg train car moving at a speed of 500 km/h.

54. In problem 49, estimate the rate of change of momentum (magnitude) for a 55,000 kg train car, as it accelerates. What is the connection between this rate and force?

55. Studies show that the cost of launching a satellite could be significantly reduced if the rocket is accelerated up a long incline before the rocket engines are fired up. The speed of the rocket must be about 950 km/h before its engines are fired up, in order to obtain the desired savings. Assume that a "launcher" with the rocket and payload has a total mass of 2000 kg. Let the launcher be accelerated along a 30° incline (in mountains), and the length of the track be 1 km.
   a) What acceleration is needed to achieve a speed of 950 km/h?
   b) What driving (external) force (parallel to the incline) is needed to provide this acceleration? (Ignore friction and air resistance)

56. In problem 55, estimate how long it would take to achieve a speed of 950 km/h.

57. In problem 55, estimate the speed of the launcher:
   a) after the first second of acceleration;
   b) after it covered the first 100 m of the track.

58. In problem 55, assume that the energy dissipated by resistive forces (air resistance, any kind of friction) is 100 MJ.
   a) How much energy is needed to accelerate the launcher to 950 km/h?

b) What must be the average power supplied by the driving force?

59. In problem 55, assume that the driving force is shut off after the first second of acceleration.

a) Find the distance traveled by the launcher (including the distance traveled during the first second) before the launcher stops momentarily. (Ignore resistive forces)

b) How long does it take to cover this distance?

60. In problem 55, assume that the driving force is shut down after the first 10 m of acceleration. Next, assume that the "resistive force" F_r in this case is similar to the kinetic friction force and can be described by the expression $F_r = \mu F_n$ where F_n is the normal force and $\mu = 0.1$. The launcher continues to move up the incline, then it stops momentarily, and then it slides all the way back down to the bottom of the incline.

a) From the moment when the driving force is shut down, how long does it take for the launcher to return to the bottom of the incline?

b) Find the distance traveled by the launcher during that time.

c) Find the magnitude of the launcher's displacement during that time.

61. Consider a tsunami generated in the deep ocean and moving toward a shore 4000 km away. Let the speed of the wave be 150 m/s and its wavelength be 700 km.

a) How long does it take tsunami to reach the shore?

b) Find the period of the wave.

c) Find the frequency of the wave.

62. Consider the front wall of a tsunami, of height 20 m and area 20 km $\times$ 100 m, hitting the shore at 10 m/s. Take the density of seawater to be 1025 kg/m^3.

a) Find the kinetic energy of the front wall.

b) Find the gravitational potential energy of the front wall (take the gravitational potential energy to be zero on the ground)

c) Find the momentum (magnitude) of the front wall.

63. In problem 61, imagine that a detector located near the ocean bottom has detected the change in pressure associated with the tsunami. This information is transmitted with a sound wave to a buoy on the ocean surface. Then the information is transmitted with an electromagnetic wave to a satellite and from the satellite to the shore. Let the detector be 2 km beneath the buoy, the distance between the satellite and the buoy be 1000 km, and the distance between the satellite and the shore station be 5000 km. How long does it take to transfer information from the detector to the shore station? Take the speed of the electromagnetic wave to be the speed of light 3×10^8 m/s, and the speed of the sound wave in the water to be 1.4 km/s.

64. It was found that low-frequency sound (below 1 kHz) does not attenuate quickly in water. A modern low-frequency active sonar (LFA) emits sound pulses in the frequency range 100–500 Hz. Assume that the frequency of a particular pulse is 200 Hz, and its duration is 10 s. Taking the speed of sound in the water as 1.4 km/s, estimate:
   a) the wavelength of the sound wave;
   b) the period of the sound wave;
   c) the length of the sound pulse in the water;
   d) the number of wavelengths in the sound pulse.

65. A tau lepton τ^- decays into a muon μ^-, a muon antineutrino $\bar{\nu}_\mu$, and a tau neutrino ν_τ. The angular momentum components along the x-axis are the following:

$$L_x = \frac{\hbar}{2}, \quad \text{for} \quad \tau^-, \tag{5.70}$$

$$L_x = \frac{\hbar}{2}, \quad \text{for} \quad \mu^-, \tag{5.71}$$

$$L_x = \frac{\hbar}{2}, \quad \text{for} \quad \nu_\tau, \tag{5.72}$$

where $\hbar = h/2\pi$, and h is Planck's constant. What is the value of L_x for the muon antineutrino $\bar{\nu}_\mu$?

66. In problem 65 let the momentum of the muon $\vec{\mathbf{p}}(\mu^-)$ be equal to momentum of the tau lepton $\vec{\mathbf{p}}(\tau^-)$. What then is the relationship between the momenta of the two neutrinos?

Chapter 6

ELECTRICITY AND MAGNETISM

6.1. Review

1. An atom consists of a nucleus and electrons, and the nucleus consists of protons and neutrons. The interaction which is responsible for the attraction between the nucleus and electrons is the electric interaction, which is a special case of the electromagnetic interaction.

2. To describe the electric interaction one assigns to an electron a negative electric charge of $-e$, to a proton a positive electric charge of e and to a neutron zero electric charge.

3. If the difference between the numbers of protons N_p and electrons N_e in a "common" macroscopic object is large, one says that the macroscopic object is electrically charged, i.e. it has a macroscopic electric charge

$$q = e(N_p - N_e) \,. \tag{6.1}$$

The macroscopic electric charge can be positive (if $N_p > N_e$) or negative (if $N_p < N_e$).

4. "Coulomb's Law" states that the electric force exerted by electrically charged particle 1 on electrically charged particle 2 is given by

$$\vec{\mathbf{F}}_{12} = k_c \frac{q_1 q_2}{r_{12}^2} \vec{\mathbf{n}}_{12} \,, \tag{6.2}$$

where k_c is Coulomb's constant, r_{12} is the distance between the two charged particles, q_1 and q_2 are their electric charges and $\vec{n}_{12}$ is the unit vector which points from particle 1 to particle 2. If the electric charges q_1 and q_2 have the same sign they repel each other, otherwise they attract each other.

5. The electric potential energy for two charged particles is given by

$$U = k_c \frac{q_1 q_2}{r_{12}} + \text{constant} \qquad (6.3)$$

6. The electric field produced by charged particle 1 at an arbitrary point P is given by

$$\vec{E}_P = k_c \frac{q_1}{r_{1P}^2} \vec{n}_{1P} , \qquad (6.4)$$

where r_{1P} is the distance between charged particle 1 and the point P, and $\vec{n}_{1P}$ is the unit vector which points from particle 1 to the point P. The electric field of a positive charge points out of the charge and the electric field of the negative charge points into the charge.

7. The electric field produced by a system of particles at the point P is the vector sum of all the electric fields produced by every charged particle (the "Superposition Principle"):

$$\vec{E}_P = \sum_n \vec{E}_{nP} = k_c \sum_n \frac{q_n}{r_{nP}^2} \vec{n}_{nP} . \qquad (6.5)$$

If an additional particle of charge q is placed at the point P, the electric force exerted by all the other charged particles on it is given by

$$\vec{F}_P = q \vec{E}_P . \qquad (6.6)$$

For a positively charged particle, the direction of the electric force $\vec{F}_P$ is the same as the direction of the electric field $\vec{E}_P$. For a negatively charged particle, $\vec{F}_P$ is directed opposite to $\vec{E}_P$.

8. The electric flux through a surface S is defined as

$$\Phi_e = \int_S \vec{\mathbf{E}} \cdot d\vec{\mathbf{A}}.$$ (6.7)

The vector $d\vec{\mathbf{A}}$ is perpendicular to the surface S at any point on the surface, and its magnitude dA is equal to an infinitesimally small area around that point. For a closed surface $d\vec{\mathbf{A}}$ points out of the enclosed volume, and for an open surface one can choose either of the two possible directions.

9. The electric flux through a closed surface is connected to the net charge inside the surface ("Gauss's Law"):

$$\Phi_e = \sum_n \frac{q_n}{\epsilon_o},$$ (6.8)

where $\epsilon_o = 1/4\pi k_c$ is called the permittivity of free space.

10. The electric field can be visualized using electric field lines. For an electric field produced by electrically charged particles, electric field lines start from the positively charged particles or from infinity and end on negatively charged particles or at infinity. The density of the electric field lines is proportional to the magnitude of the electric field. At any point on an electric field line, the electric field is tangential to the line at that point.

11. Let a system of fixed charged particles create an electric field $\vec{\mathbf{E}} = \vec{\mathbf{E}}(\vec{\mathbf{r}})$. If an additional charged particle moves from the point P_i to the point P_f the work done by the electric force on the particle is given by

$$W = +q \int_{P_i}^{P_f} \vec{\mathbf{E}} \cdot d\vec{\mathbf{r}},$$ (6.9)

where q is the charge of the additional particle.

12. The quantity

$$V_P = -\int_{P_o}^{P} \vec{\mathbf{E}} \cdot d\vec{\mathbf{r}}$$ (6.10)

is called the electric potential at point P relative to point P_o, where P_o is an arbitrarily chosen point. The electric potential energy of an additional charged particle of charge q placed at the point P is given by

$$U_P = qV_P + \text{constant} \tag{6.11}$$

13. The electric potential produced by charged particle 1 at an arbitrary point P relative to "infinity" is given by

$$V_P = k_c \frac{q_1}{r_{1P}} . \tag{6.12}$$

The electric potential produced by a system of charged particles equals the sum of the individual electric potentials:

$$V_P = \sum_n V_{nP} = k_c \sum_n \frac{q_n}{r_{nP}} . \tag{6.13}$$

14. A conventional conductor contains a huge number of charged particles, which can move inside the conductor under the action of a macroscopic electric field. It is customary to consider free positively charged particles inside the conductor, while typically the free particles are actually negatively charged electrons. For example, in copper there are more than 10^{22} mobile electrons per cubic centimeter (cm^{-3}). The results of the analysis do not depend on the sign of free charges.

15. Under conditions of electrostatic equilibrium, the macroscopic electric field inside a good conductor is zero, and all points in the conductor have the same electric potential. If a conductor has a macroscopic electric charge this charge is distributed on the conductor's surface. The electric field near the conductor's surface is perpendicular to the surface and its magnitude is

$$E = \frac{\sigma}{\epsilon_o} , \tag{6.14}$$

where σ is the local electric surface charge density.

16. The simplest capacitor consists of two conductors separated by a dielectric (insulator). The conductors carry the same magnitude

but opposite sign electric charges. The charge on the capacitor Q is defined as the magnitude of the charge on either conductor, and the voltage on the capacitor V is the absolute value of electric potential difference between the two conductors. The value of the stored charge Q is proportional to the voltage V:

$$Q = CV.$$ (6.15)

The coefficient C is called the capacitance of the capacitor.

17. For a parallel-plate capacitor the capacitance is given by

$$C = \frac{\epsilon A}{s},$$ (6.16)

where ϵ is the permittivity of the insulating substance which fills the space between the conducting plates, A is the area of the plates and s is the separation between the plates. In the special case where the insulator is vacuum (i.e. empty space), $\epsilon = \epsilon_o$, the permittivity of free space.

18. The electric potential energy of a charged capacitor (with accuracy to a constant) is given by

$$U = \frac{QV}{2} = \frac{CV^2}{2} = \frac{Q^2}{2C}.$$ (6.17)

19. A battery is a device which separates positively and negatively charged particles, producing a permanent electric potential difference between its terminals. The magnitude of this potential difference is called the "terminal voltage" or more often simply the voltage. If a battery is connected to a capacitor, the voltage on the capacitor will quickly become equal to the terminal voltage.

20. If two capacitors are connected in series, the electric charge stored by the first capacitor is equal to the charge stored by the second capacitor and equal to the total amount of charge stored by the two capacitors. The voltage across the two capacitors equals the sum of the individual voltages. Thus, the effective capacitance C of two

capacitors is given by

$$\frac{1}{C} = \frac{1}{C_1} + \frac{1}{C_2}. \tag{6.18}$$

21. If two capacitors are connected in parallel, the voltage across the first capacitor is equal to the voltage across the second capacitor and equal to the total voltage across the two capacitors. The total electric charge stored by the two capacitors equals the sum of the individual charges. Thus, the effective capacitance of the combination, C, is given by

$$C = C_1 + C_2. \tag{6.19}$$

22. The simplest direct current (DC) electric circuit consists of a battery connected to a "modestly conducting resistor" [a] or simply a resistor. The analysis of a DC circuit does not depend on which charge sign (positive or negative) is free to move around the circuit. It is customary to consider moving "effective positive charge", while, in fact, normally it is the negatively charged electrons that are free to move. Inside the battery in a DC circuit, an "effective chemical force" constantly "pushes" effective positive charge from the negative terminal to the positive terminal, supporting a constant value of the external terminal voltage. The battery terminals produce a constant electric field inside the modestly conducting resistor. The electric force acting on the effective positive charge inside the resistor "pushes" the charge from the positive terminal to the negative terminal through the resistor. In the steady state, the chemical force in the battery is balanced by the electric force produced by the battery terminals, and the electric force in the resistor is balanced by an "effective friction force". Thus, the amount of electric charge passing through any cross-sectional area per unit time remains constant.

23. The work needed for the chemical force to move a unit positive charge from the negative terminal of the battery to the positive

---

[a]The term modestly conducting resistor is used instead of conductor because the current through a good conductor can be large enough to damage both the conductor and the battery. The modestly conducting resistor could consist of, for example, a pair of conducting wires and an electric light bulb.

terminal is called the EMF. EMF stands for electromotive force, but it is rather "electromotive potential". In an ideal battery the terminal voltage equals the EMF. In a real battery connected to a modestly conducting resistor, the terminal voltage is smaller than EMF.

24. The electric current in a DC electric circuit is defined as the magnitude of the electric charge passing through a cross-sectional area of a circuit in a unit time. Electric current is a positive scalar, it does not have a direction. However, it is customary to use the term "direction of the current", which means the direction of motion of effective positive charge.

25. For a resistor, the electric current I is proportional to the magnitude of electric potential difference V (voltage) across the resistor ("Ohm's Law"):

$$I = \frac{V}{R}.\tag{6.20}$$

The quantity R is called the resistance of the resistor. For a resistor of fixed cross-sectional area A and length L, the resistance is proportional to L and inversely proportional to A:

$$R = \frac{\rho L}{A}.\tag{6.21}$$

The quantity ρ is called the resistivity of a substance.

26. One can describe the energy associated with the electric current in a DC electric circuit. This energy is dissipated due to the effective friction in the resistor. (Typically the energy of an electric current is transformed into internal (thermal) energy in the resistor, and then into electromagnetic radiation.) The magnitude of the rate of dissipation of electric energy (the dissipated power) is given by

$$P = IV = I^2 R = \frac{V^2}{R}.\tag{6.22}$$

Here I is the current in the circuit and V is the voltage across the resistor, of resistance R. The internal (chemical) energy of the battery is continuously transformed into the electric energy to balance the dissipated energy.

27. If two resistors are connected in series, the voltage across the two resistors equals the sum of the individual voltages. The current through the first resistor equals the current through the second one and equals the total current in the DC circuit. Thus, the effective resistance of the two resistors is

$$R = R_1 + R_2 \,. \tag{6.23}$$

28. If two resistors are connected in parallel, the total current in the DC circuit equals the sum of the individual currents. The total voltage equals the voltage on either resistor (both resistors have the same voltage). Thus,

$$\frac{1}{R} = \frac{1}{R_1} + \frac{1}{R_2} \,. \tag{6.24}$$

29. If electric currents flow through two conductors, there is a magnetic interaction between these conductors. To describe this interaction one introduces the magnetic field $\vec{\mathbf{B}}$. Every conductor carrying current produces a magnetic field, which acts on any other conductor carrying a current.

30. If a current I flows through a linear conducting segment $d\vec{\mathbf{L}}$, the magnetic field $d\vec{\mathbf{B}}$ produced by this segment at the point P is given by (the "Biot–Savart Law")

$$d\vec{\mathbf{B}}_P = \frac{\mu_o}{4\pi} \frac{I d\vec{\mathbf{L}} \times \vec{\mathbf{n}}_P}{r_P^2} \,. \tag{6.25}$$

Here μ_o is the permeability of free space, the vector $d\vec{\mathbf{L}}$ shows the direction of flow of the effective positive charge, dL is the length of the infinitesimally short segment, r_P is the distance between the segment and the point P, and $\vec{\mathbf{n}}_P$ is the unit vector which points from the segment $d\vec{\mathbf{L}}$ to the point P.

31. The total magnetic field produced by a thin conducting wire

at the point P can be found using

$$\vec{\mathbf{B}}_P = \int_L d\vec{\mathbf{B}}_P = \frac{\mu_o I}{4\pi} \int_L \frac{d\vec{\mathbf{L}} \times \vec{\mathbf{n}}_P}{r_P^2} . \qquad (6.26)$$

where the integral is taken over the whole wire.

32. The magnetic moment $\vec{\mathbf{m}}$ of a planar current loop is defined as

$$\vec{\mathbf{m}} = \vec{\mathbf{n}} I A . \qquad (6.27)$$

Here I is the current in the loop, A is its area, and $\vec{\mathbf{n}}$ is a unit vector, which is perpendicular to the plane of the loop. The direction of motion of the effective positive charge in the loop is counterclockwise relative to the direction of $\vec{\mathbf{n}}$.

33. The magnetic field produced by a small planar current loop at a point P far from the loop is given by

$$\vec{\mathbf{B}}_P = \frac{\mu_o}{4\pi} \frac{3(\vec{\mathbf{m}} \cdot \vec{\mathbf{n}}_P)\vec{\mathbf{n}}_P - \vec{\mathbf{m}}}{r_P^3} . \qquad (6.28)$$

Here $\vec{\mathbf{n}}_P$ is the unit vector which points from the loop to the point P, and r_P is the distance between the loop and the point P.

34. To describe the magnetic properties of a particle one introduces the magnetic moment of the particle $\vec{\mathbf{m}}$. The magnetic field produced by the particle at a distant point P is given by the same expression as the magnetic field produced by a small current loop. For a spherical magnet this formula is valid for any point P outside the magnet.

35. The magnetic field, like an electric field, can be visualized using magnetic field lines. Unlike the electric field produced by fixed electrically charged particles, the magnetic field is a "curl" field. The magnetic field lines are closed curves: they do not have starting or end points.

36. Ampere's Law provides an important integral expression for the

magnetic field produced by an electric current

$$\oint_L \vec{\mathbf{B}} \cdot d\vec{\mathbf{L}} = \mu_o I \,, \tag{6.29}$$

where the integral $\oint$ is taken over a closed line L, and current I is the net current which passes through the surface bounded by the closed line. The direction of $d\vec{\mathbf{L}}$ is counterclockwise relative to the direction of the net current.

37. If an electric field changes with time, it generates a magnetic field. The extended form of the Ampere's Law takes this source of magnetic field into account:

$$\oint_L \vec{\mathbf{B}} \cdot d\vec{\mathbf{L}} = \mu_o \left(I + \epsilon_o \frac{d\Phi_e}{dt} \right) \,. \tag{6.30}$$

The second term on the right-hand side of this expression is often called the "displacement current." The vector $d\vec{\mathbf{L}}$ is counterclockwise relative to the vector $d\vec{\mathbf{A}}$ in the expression for the electric flux Φ_e (6.7).

38. The magnetic force produced by a magnetic field $\vec{\mathbf{B}}$ on a particle with electric charge q, moving with velocity $\vec{\mathbf{v}}$ (the "Lorentz Force") is given by

$$\vec{\mathbf{F}} = q\vec{\mathbf{v}} \times \vec{\mathbf{B}} \,. \tag{6.31}$$

39. The magnetic force produced by a magnetic field $\vec{\mathbf{B}}$ on a linear electric current element $I d\vec{\mathbf{L}}$ is given by

$$d\vec{\mathbf{F}} = I d\vec{\mathbf{L}} \times \vec{\mathbf{B}} \,. \tag{6.32}$$

If a planar current loop with magnetic moment $\vec{\mathbf{m}}$ is placed into a uniform magnetic field, the net magnetic force on the loop is zero. The torque produced by the magnetic field is given by

$$\vec{\tau} = \vec{\mathbf{m}} \times \vec{\mathbf{B}} \,. \tag{6.33}$$

The same expression is valid for a magnetic particle of magnetic moment $\vec{\mathbf{m}}$.

41. For both a current loop and a magnetic particle, one can introduce the magnetic potential energy associated with the direction of the magnetic moment $\vec{\mathbf{m}}$ relative to the direction of a uniform magnetic field $\vec{\mathbf{B}}$:

$$U = -\vec{\mathbf{m}} \cdot \vec{\mathbf{B}} + \text{constant} \qquad (6.34)$$

42. The magnetic flux through a surface S is defined just like the electric flux as

$$\Phi_m = \int_S \vec{\mathbf{B}} \cdot d\vec{\mathbf{A}} . \qquad (6.35)$$

43. If the magnetic flux changes with time it generates a "curl" electric field. The electric field lines in this case are closed: they do not have starting or end points.

44. "Faraday's Law of Induction" provides an integral expression relating the "curl" electric field and the magnetic flux through a surface bounded by a closed line L

$$\oint_L \vec{\mathbf{E}} \cdot d\vec{\mathbf{L}} = -\frac{d\Phi_m}{dt} , \qquad (6.36)$$

where the integral is taken over the closed line L and the direction of $d\vec{\mathbf{L}}$ is counter clockwise relative to the direction of $d\vec{\mathbf{A}}$ in the expression for the magnetic flux Φ_m (6.35).

45. Let the magnetic flux through a surface bounded by a conducting loop change, due to a change in the magnetic field or due to motion of the conducting loop. In both cases, the changing magnetic flux induces an electromotive force $\mathcal{E}$ in the loop

$$\mathcal{E} = -\frac{d\Phi_m}{dt}, \qquad (6.37)$$

and an induced electric current $I = |\mathcal{E}|/R$, where R is the loop resistance. An overall positive sign for $\mathcal{E}$ means that the direction of the induced current is counterclockwise relative to the direction of $d\vec{\mathbf{A}}$ in the expression for the magnetic flux Φ_m, an overall negative sign for $\mathcal{E}$ corresponds to the opposite direction of induced current.

46. "Lenz's Law" provides an independent way to find the direction of the induced current. It states that the magnetic field of the induced current always opposes the change in the magnetic flux through the conducting loop.

47. The magnetic flux produced by a current loop (or coil) through a surface bounded by this loop is proportional to the current

$$\Phi_m = L_m I \,. \tag{6.38}$$

The coefficient L_m is called the inductance of the loop.

48. The magnetic energy stored in a current loop (or coil) is given by

$$E_m = \frac{1}{2} L_m I^2 \,. \tag{6.39}$$

49. Electromagnetic radiation is an electromagnetic wave consisting of propagating disturbances of the electric and magnetic fields, (i.e. electromagnetic fields) in free space or in matter. If, in a region of space the electric and magnetic fields depend only on one spatial co-ordinate (say x) and time, the electromagnetic wave in that region is called a plane electromagnetic wave. If the electric fields oscillates in a single plane, the wave is said to be linearly polarized, and the polarization direction is taken to be along the direction of oscillation of the electric field.

50. The simplest traveling electromagnetic wave is a plane, linear polarized harmonic wave, in which the electric field oscillates along one axis (say, the y-axis) the magnetic field oscillates along an-other axis (say z-axis), and the wave propagates along the third axis (x-axis). We consider only this simplest case below.

51. The electric and magnetic fields in the simplest electromagnetic wave are given by

$$E_y = E_o \cos(kx - \omega t + \phi), \quad B_z = B_o \cos(kx - \omega t + \phi)\,, \tag{6.40}$$

if the wave travels in the positive x-direction and

$$E_y = E_o \cos(kx + \omega t + \phi), \quad B_z = -B_o \cos(kx + \omega t + \phi)\,, \tag{6.41}$$

if the wave travels in the negative x-direction. Here k is the wave number, ω is the angular frequency, and ϕ is the phase constant of the wave.

52. The speed of an electromagnetic wave is commonly designated as c:

$$c = \frac{\omega}{k} = \frac{\lambda}{T} = \lambda f \,, \tag{6.42}$$

where λ is the wavelength, T is the period and f is the frequency of the wave. In free space the value of c is related to the permittivity ϵ_o and the permeability μ_o:

$$c = (\epsilon_o \mu_o)^{-\frac{1}{2}} \,. \tag{6.43}$$

The ratio of the amplitudes of the electric and magnetic fields in an electromagnetic wave is equal to the speed c:

$$\frac{E_o}{B_o} = c \,. \tag{6.44}$$

53. The energy density associated with the electric field $\vec{\mathbf{E}}$ and the magnetic field $\vec{\mathbf{B}}$ are given by

$$u_E = \frac{\epsilon_o E^2}{2}, \quad u_B = \frac{B^2}{2\mu_o} \,. \tag{6.45}$$

54. The average energy density (over a period) in a region of electromagnetic fields is

$$u_{av} = \frac{E_o B_o}{2\mu_o c} \,. \tag{6.46}$$

55. The power per unit area carried by an electromagnetic wave through a perpendicular surface can be described by the magnitude of the "Poynting vector" $\vec{\mathbf{S}}$

$$\vec{\mathbf{S}} = \frac{\vec{\mathbf{E}} \times \vec{\mathbf{B}}}{\mu_o} \,. \tag{6.47}$$

The direction of the Poynting vector is the direction of propagation of the wave.

56. The intensity of an electromagnetic wave, (i.e. electromagnetic radiation) is the average (over a period) of the Poynting vector magnitude

$$I_r = \frac{E_o B_o}{2\mu_o} . \tag{6.48}$$

57. The radiation pressure P_r is the average force per unit area (magnitude) exerted by an electromagnetic wave incident on an object. If the wave is incident perpendicular to an absorbing surface, the radiation pressure is given by

$$P_r = \frac{I_r}{c} . \tag{6.49}$$

58. If an electromagnetic wave is incident on an object whose size is much greater than the wavelength, one can use the "ray approximation." This means the waves propagate along straight lines (called rays), which can change their direction only at the boundary between two substances. The ray approximation is widely applied for the visible part of the electromagnetic spectrum ("ordinary visible light").

59. If light is incident on a smooth flat surface, the angle of incidence is equal to the angle of reflection, where the angles are measured from the normal to the surface (the "Law of Reflection," see Fig. 8).

60. If light passes through the boundary between two substances the angle of refraction θ_2 is connected to the angle of incidence θ_1 by "Snell's law" (see Fig. 9):

$$n_1 \sin \theta_1 = n_2 \sin \theta_2 , \tag{6.50}$$

where n_1 is the refractive index of the first substance, n_2 is the refractive index of the second substance. The refractive index of a substance is given by $n_i = c/v_i$ where c and v_i are the speeds of light in free space and in the substance of interest respectively.

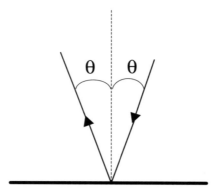

Fig. 8. Reflection of light in the ray approximation.

61. If a linearly polarized electromagnetic wave of intensity I_r passes through a polarizing sheet, the intensity of the transmitted light I_r' is given by "Malus's Law":

$$I_r' = I_r \cos^2 \theta, \qquad (6.51)$$

where θ is the angle between the electric field of the incident electromagnetic wave and the transmission axis of the polarizing sheet. For the transmitted light, the electric field oscillates along the transmission axis of the polarizing sheet, and the magnetic field is perpendicular to the electric field.

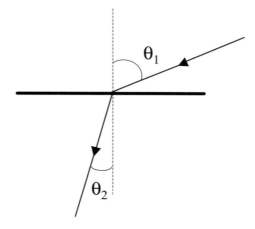

Fig. 9. Refraction of light in the ray approximation.

6.2.   Problems

1. Estimate the magnitude of the electric field produced by an up quark at a distance of 2 fm from the quark. Use the data in Table 2 in Chapter 2.

2. Estimate the electric potential produced by an up quark at a distance of 2 fm from the quark, assuming zero electric potential at infinity.

3. Estimate the magnitude of the electric force between up and down quarks if the distance between them is 1 fm.

4. A tau lepton was created in a uniform electric field of magnitude 1 MV/m. Estimate the mean distance traveled by the tau lepton during its lifetime if its initial velocity is 10 km/s pointing in the direction of the electric field. Take the mean lifetime of the tau lepton to be 290 fs. Use the data in Table 1 in Chapter 2.

5. Solve problem 4 if the initial velocity points in the direction opposite to the direction of the electric field.

6. A tau lepton was created between the plates of a parallel-plate capacitor of area 1 cm^2 and separation 1 mm. Estimate the magnitude of the tau lepton's acceleration if the charge on the capacitor is 1 μC. What is the direction of the acceleration relative to the plates?

7. A muon was created in a region with a uniform magnetic field of magnitude 10 T. If the initial velocity of the muon is 100 km/s perpendicular to the direction of the magnetic field, find the magnitude of its acceleration. Use the data in Table 1 in Chapter 2.

8. In problem 7 find the mean distance traveled by the muon during its lifetime. Take the mean lifetime of the muon to be 2.2 μs.

9. Estimate the electric potential energy of the three quarks in a proton. (Take the potential energy to be zero when the distance be-

tween the quarks is infinite). Assume that the quarks are stationary and located at the vertices of an equilateral triangle of side 1.2 fm and use Tables 2 and 3 in Chapter 2 for the other data needed.

10. Solve problem 9 for a neutron.

11. Find the magnitude of the acceleration of a tau lepton located at a distance of 1 nm from a proton, assuming that the only force acting on the tau lepton is the electric force produced by the proton.

12. Solve problem 11 for a muon.

13. Estimate the magnitude of the electric field produced by a ^3_2He nucleus at a distance of 1 nm from its center.

14. In problem 13 estimate the electric potential produced by the nucleus. Take zero potential at an infinite distance from the nucleus.

15. Estimate the maximum possible electric force (magnitude) between two stable nuclei at a distance of 1 nm between their centers. ($^{209}_{83}Bi$ is the stable isotope with the largest electric charge.)

16. Estimate the magnetic potential energy difference for two states of a ^3_2He nucleus if: (1) the magnetic moment of the nucleus points in the direction of a magnetic field of magnitude 1 T, and (2) the magnetic moment of the nucleus points in the direction opposite to the direction of the magnetic field. (The magnetic moment of ^3_2He is 2.1 μ_n.)

17. In problem 16 find the maximum possible torque (magnitude) produced by the magnetic field on the nucleus.

18. A $^{209}_{83}Bi$ nucleus is placed at the origin. Its magnetic moment ($m = 4\mu_B$) points in the positive z-direction. Find the magnitude and direction of the magnetic field produced by the nucleus at the point $x = y = 0, z = 1$ nm.

19. In problem 18 find the magnitude and direction of the magnetic field at points
   a) $x = y = 0, z = -1$ nm;
   b) $y = z = 0, x = 1$ nm.

20. Deuterium and tritium can fuse producing helium. Estimate the force of electric repulsion (magnitude) between deuterium ${}^2_1H$ and tritium ${}^3_1H$ nuclei with a distance of 10 fm between their centers.

21. In problem 20, how much energy is needed to provide a distance of 10 fm between the two nuclei if the initial distance was very much greater than 10 fm.

22. In field emission transmission electron microscopy (FETEM) electrons are drawn out of a cathode by a high voltage. In a modern device the electric potential difference can be as high as 1 MV. Assuming that the initial speed of the electrons is small compare to the final speed, estimate the final kinetic energy of the electrons (in joules) accelerated by the 1 MV voltage.

23. To compress nuclei and produce a new form of nuclear matter, scientists smash a high energy beam of lead ions into a fix target of lead or gold atoms. After a collision the momentary nuclear matter density is estimated to be 20 times the normal nuclear density. Assume that two lead nuclei collide, producing a "fireball" of 20 times normal nuclear density. Estimate the radius of the fireball in terms of the radius of a lead nucleus, and the magnitude of the electric field near the fireball in terms of the magnitude of the electric field near the lead nucleus. Assume that the mass of the fireball is approximately double the mass of the lead nucleus.

24. In one of the versions of magnetic resonance force microscopy (MRFM) a cantilever with a small magnet on its tip oscillates near a sample causing cyclic change in the direction of the magnetic moment of an atom in the sample. In turn, the atomic magnetic moment reversals cause a change in the cantilever oscillation frequency, which can be detected. Assume that both the magnetic moment of the mag-

Fig. 10. A Halbach array. The arrows show the directions of the individual magnetic moments.

net and the atomic magnetic moment in the sample point along the line connecting them. Estimate the magnetic field (magnitude) produced by the magnet at the atom and magnetic field produced by the atom at the magnet, if they are separated by the distance 1 μm. Take the magnetic moment of the magnet to be 1.5×10^{-12} Am^2, and the magnetic moment of the atom to be 9.3×10^{-24} Am^2.

25. In problem 24 assume that the cantilever oscillates (vibrates) with an amplitude of 10 nm. Estimate the range of the change in magnetic field at the atom.

26. The intrinsic magnetic moment of the muon m_μ has been measured recently with an accuracy of about 1.3 parts per million. The precise value of m_μ is extremely important for verification of the theory of fundamental elementary particles. Assume that the measured magnetic moment of the muon is greater than the theoretically predicted value by 2 parts per million. Let a muon be placed in a high magnetic field with $B = 20$ T.

a) Estimate the difference between the theoretical and experimental values of intrinsic magnetic moment (take the theoretical value to be 9.3×10^{-24} J/T.

b) Estimate the corresponding shift in the value of the difference between the maximum and minimum magnetic energy of the muon.

27. A "Halbach Array" of permanent magnets is shown in Fig. 10.

The arrows in the figure show the direction of the magnetic moment (from south pole to north pole). At what sides of the Halbach array (top or bottom) do the magnetic fields from neighboring magnets reinforce each other? On what side do they cancel each other?

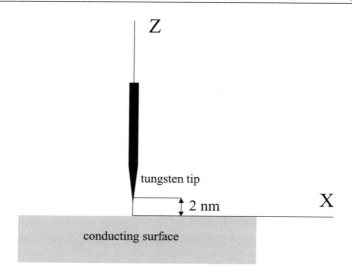

Fig. 11.   A tungsten tip near a conducting surface.

28. In a scanning tunneling microscope a tungsten tip is brought very close to the conducting surface under investigation. A small tunneling current of about 2 pA flows from the tip to the surface (electrons "jump" from the surface to the tip). Consider a system of coordinates with the origin on the conducting surface and z-axis directed along the tip (see Fig. 11).

Find the magnetic field (in unit vector notation) produced by the current between the surface and the tip at the point P ($x_P = 36$ nm, $y_P = 64$ nm, $z_P = 0$). Take the separation between the tip and the surface to be 2 nm.

29. In problem 28 consider a paramagnetic atom at the point P. The magnetic moment of the atom has a magnitude of one Bohr magneton (μ_B) and points in the positive x-direction.

a) Find the torque (in unit vector notation) produced by the magnetic field of the tunneling current on the atom.

b) Find the change of the atomic magnetic energy due to the interaction with the tunneling current, if the direction of the atomic magnetic moment reverses.

30. In problem 28 consider the probe and the sample as a capacitor with the capacitance of 10^{-16} F. If the voltage between the tip and sample is 1 V, find the induced tip charge.

31. It is estimated that conducting nanotubes can withstand electric current densities of up to 10^{13} A/m^2, 1000 times greater than copper. Consider a conducted wire made of carbon nanotubes. Let the current density in the wire $j = 1$ TA/m^2 (T here stands for tera = 10^{12}). If the cross-sectional area of the wire is 1 mm^2 estimate the charge which passes through the cross-sectional area in 10 seconds.

32. Consider a small 1 cm segment of the wire discussed in problem 31, located at the origin of coordinates. If the current flows in the positive z-direction find (in unit vector notation):

a) the magnetic field produced by the segment at the point P (10 cm, 15 cm, 20 cm);

b) the magnetic force produced by the segment on an electron at the point P, moving at 10^5 m/s in the positive z-direction.

33. Consider a circular current loop of radius 1 cm made of the wire discussed in problem 31.

a) What is the magnetic moment (magnitude) of the current loop?

b) If the current loop is placed at the origin, and the current flows in the x–y plane in a counterclockwise direction relative to the z-axis, find the magnetic field at the point P (10 cm, 15 cm, 20 cm) in unit vector notation.

c) Find the torque on an atom (in unit vector notation) at the point P if the atom has a magnetic moment of two Bohr magnetons, which points in the positive z-direction.

34. Consider a square current loop of side 1 cm made of the material described in problem 31, which is placed in a uniform magnetic field of 1 T. The plane of the loop is perpendicular to the magnetic field.

a) What is the change of the magnetic energy of the loop (absolute value) if the direction of the current reverses?

b) What is the torque (magnitude) on the loop produced by the

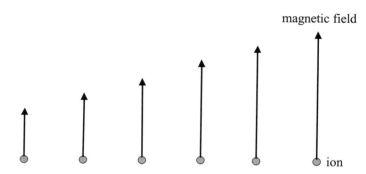

Fig. 12.   A nonuniform magnetic field on a chain of ions.

magnetic field?

35. One of the proposals for quantum computation suggests using a chain of tellurium ions on a silicon surface. To allow manipulations with single ions, the sample is placed in a nonuniform magnetic field. Assume that the magnetic field (magnitude) on the left end of the chain is 5 T. The magnetic field for each ion is 5×10^{-4} T greater than the magnetic field at its left neighbor. (See Fig. 12, where gray dots show tellurium ions and arrows show the magnetic field on the ions.)

Let the chain contain 100 ions. Consider two states of an ion: (1) the magnetic moment of the ion points in the direction of the magnetic field, (2) it points in the opposite direction. Find the difference in magnetic energy for these two states:

a) for the left end ion;

b) for its neighbor;

c) for the right end ion.

The magnetic moment of the tellurium ion equals one Bohr magneton (μ_B).

36. In problem 35 consider the same chain of tellurium ions but in a uniform magnetic field. To "select" an ion, a small spherical magnet moves along the chain and "stops" exactly above the selected ion (see Fig. 13).

The quantity $\mu_o m/V$ is called the magnetic induction of the mag-

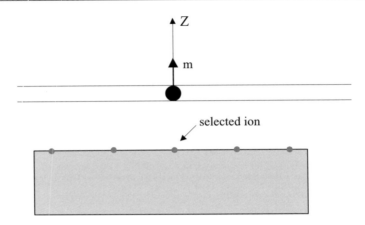

Fig. 13. A movable magnet "selects" an ion on the surface of a sample.

net, where m is the magnetic moment of the magnet, and V is its volume. Assume that the magnetic induction is 2.2 T, the radius of the magnet is 5 nm, and the distance between the center of the magnet and the selected ion is 15 nm. Find:

a) the magnetic field produced by the magnet on the selected ion;

b) the maximum possible torque (magnitude) produced by the magnet on the selected ion;

c) The maximum change in the magnetic energy of the selected ion connected to the direction of its magnetic moment.

37. One of the proposals for quantum computation suggests using a chain of complicated "molecules", which are designated as $N@C_{60}$, on a silicon surface. Such molecules can be precisely positioned on a silicon surface using scanning tunneling microscopy. C_{60} molecules can be considered as small (≈ 1 nm) traps for paramagnetic nitrogen atoms. To create a nonuniform magnetic field at the nitrogen atoms, it was suggested that so-called "micro-patterned wires" be used. Assume that two wires are placed on the silicon surface as shown in Fig. 14.

Assume that an electric current of 1 A flows in both wires in the positive y-direction (into the page), the distance between the wires $\Delta = 1$ μm, and the diameter of a wire $d = 1$ μm. Find the magnetic field (in unit vector notation) produced by the wires:

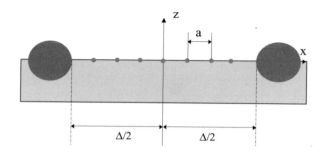

Fig. 14.   The large circles represent the micro-pattern wires (cross-sections) on the surface of the sample. The small dots represent the chain of molecules.

a) at the origin (i.e. the midpoint between the wires);
b) at the nitrogen atom at $x = a$ (take the distance between the neighboring nitrogen atoms $a = 1.14$ nm);
c) at $x = -a$.

38. In problem 37 consider three neighboring nitrogen atoms at points $x = 0, x = a$, and $x = 2a$. Find the difference in the magnetic energy for these atoms assuming that all three magnetic moments point in the positive z-direction. Take the magnetic moment of the nitrogen atom to be $3\mu_B$.

39. The luminosity (the radiated light power) of the Sun is about 4×10^{26} W. Find the energy emitted by the Sun in 1 year.

40. In problem 39 find the intensity of the Sun's electromagnetic radiation at a distance of 0.1 AU from the center of the Sun (1 AU $\approx 1.5 \times 10^{11}$ m).

41. The asteroid Aten has a semimajor axis of 0.97 AU. Estimate the radiation pressure on the surface of the asteroid when the distance between the asteroid and the center of the Sun is 0.9 AU. Consider the part of the surface which is perpendicular to the incident light. The light power radiated by the Sun is 4×10^{26} W. Assume that Aten's surface absorbs all the light falling on it.

42. Space-based x-ray observations hint at the existence of magne-tars (neutron stars which produce magnetic fields of up to 10^{11} T). Assume that an electron is traveling at 100 km/s near the surface of a magnetar. What is the maximum magnetic force (magnitude) that can be exerted on the electron, if the magnetic field is 10^{11} T.

43. In problem 42 what maximum torque (magnitude) can be exerted on the intrinsic magnetic moment of the electron, which is equal to one Bohr magneton.

44. It was shown that pulses of hard x-ray radiation can be trapped in a crystal resonator: a pair of thin slabs made from silicon and placed about 15 cm apart. Let a 10^{-10} s x-ray pulse of frequency about 4×10^{18} Hz travel back and forth between the slabs. Find:
   a) the wavelength of the x-ray;
   b) the period of the x-ray;
   c) the length of the pulse;
   d) the number of wavelengths in the pulse.

45. In problem 44 assume that after every reflection the energy of the pulse decreases by 30%.
   (a) How many reflections does the pulse experience before it loses more than 90% of its initial energy?
   (b) Estimate the distance traveled by the pulse before it loses more than 90% of its energy.
   (c) How long does it take the pulse to cover the distance found in part (b)?

46. Optical standing waves between two mirrors can trap neutral atoms at the antinodes of the wave. If the frequency of the light is 3.5×10^{14} Hz what is the distance between two neighboring traps?

47. In experiments on "stopping light" scientists irradiated a Rubidium–Helium gas mixture with a "control beam," which changes the properties of the gas. In the presence of the control beam they sent a "signal light pulse" into the gas from another source of light whose duration is τ. Inside the gas the speed of the signal pulse drops

to approximately 900 m/s. When the signal was "compressed" inside the gas, the control beam was turned off. The light inside the gas (both "control" and "signal") disappeared. When the scientists irradiated the gas with the control beam again, the gas emitted a signal light pulse of the same duration τ. In other words the signal pulse was essentially "stopped and then released" by the gas.

a) Taking $\tau = 20$ μs, find the length of the signal pulse before entering the gas.

b) Find the same length inside the gas.

48. Our Milky Way Galaxy has two small satellite galaxies which are called the "Magellanic Clouds." The smaller of these (the "Small Magellanic Cloud") lies approximately 200,000 ly away from the Sun. The space-based Hubble Space Telescope routinely produces sharp images of stars in the Small Magellanic Cloud. Assuming that the light detected by the Hubble telescope has a wavelength 600 nm estimate the number of wavelengths between a star in the Small Magellanic Cloud and the Hubble telescope.

49. Galaxy clusters are the largest "structural elements" in the universe. Typically hot gas fills the space within a cluster. The hot gas radiates x-rays, which can be detected by "x-ray telescopes." In 2002 the most distant cluster of galaxies "3C294" was discovered by the space-based Chandra X-Ray Observatory. Its size is estimated to be 6×10^5 ly. If the angular size of the cluster is about 6×10^{-5} rad estimate how long ago the x-rays detected by Chandra were emitted by the cluster.

50. A recently created composite material, made of fiberglass and copper, has a "negative refractive index" for microwave radiation of frequency about 5 GHz. The negative refractive index in Snell's Law means that the angle of refraction is negative, i.e. the refractive wave propagates on the same side of the normal as the incident wave.

a) What is the wavelength of microwaves with frequency 5 GHz?

b) Let a point source of microwaves be placed at a distance s from a slab of a composite with refractive index $n = -1$ and width w. (See Fig. 15.)

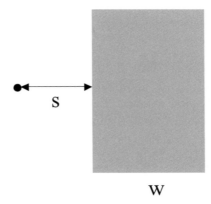

S

W

Fig. 15. A "point source" of microwaves is placed at a distance s from a slab of width w $(w > s)$.

Find the position of the image of the point source inside the slab (graphically). What is the distance between the object and the image?

c) Find the position of the final image outside the slab. What is the distance between the object and the image?

Chapter 7

HINTS TO SOLUTION

7.1. Mechanics

1. The gravitational acceleration near the ground $g = GM_e/R_e^2$. At the height h the gravitational acceleration is $g' = GM_e/(R_e + h)^2$. The relative change of gravitational acceleration in percent is

$$\frac{g - g'}{g} \times 100 \approx \frac{200h}{R_e} .$$

2. The gravitational force at a height h is given by

$$F_g = GM_em/(R_e + h)^2 = mg\left(\frac{R_e}{R_e + h}\right)^2 .$$

where m is the mass of the bacteria.

3b. $s = v\Delta t$ where v is the speed of the probe and $\Delta t = 1$ year.

4b. $F = F_g = GM_em/r_{es}^2$, where r_{es} is the distance between the center of the Earth and the spacecraft and m is mass of the spacecraft.

4c. $GM_em/r_{es}^2 = GM_om/r_{ss}^2$, where r_{ss} is the distance between the center of the Sun and spacecraft, r_{es} is the distance between the center of the Earth and the spacecraft.

5a. $s = v\Delta t$.

5b. $p = mv,\ \ K = mv^2/2$.

6a. $F_g = GM_e m/(r_e^2 + h)^2 = mg\left(\dfrac{R_e}{R_e + h}\right)^2.$

6b. $a = F/m \approx F_g/m = g\left(\dfrac{R_e}{R_e + h}\right)^2.$

7. $v = s/\Delta t.$

8. The gravitational force (magnitude) exerted by the Sun F_{gs} is $F_{gs} = GM_o m/r_{ss}^2$, where r_{ss} is the distance between the center of the Sun and the spacecraft. The gravitational force exerted by the Earth $F_{ge} = GM_e m/r_{es}^2$ where r_{es} is the distance between the center of the Earth and the spacecraft. The ratio is

$$\frac{F_{gs}}{F_{ge}} = \frac{M_o}{M_e} \cdot \left(\frac{r_{es}}{r_{ss}}\right)^2.$$

9a. Consider a system of coordinates connected to the spacecraft. For a small time interval (the time of the collision) it can be treated as an inertial system. The magnitude of the impulse I_a delivered by the spacecraft to the atom is $I_a = mv$, where v is the speed of the atom. It follows from the Action–Reaction Principle that the magnitude of the impulse delivered by the atom to the spacecraft I_s is equal to I_a. To estimate the mass of the gold atom, use the Periodic Table of Elements. The atomic mass of Au atom is approximately 197. Thus 1 mol of gold, which contains Avagadro's number of atoms has a mass of 197 g.

9b. Yes, approximately. During the small time of collision we can neglect the external gravitational forces, so the total momentum is conserved.

9c. $K = mv^2/2$. No. Kinetic energy is not conserved in inelastic collisions (in our problem we have a perfectly inelastic collision: the atom sticks to the collector).

10a. $\vec{\mathbf{L}} = \vec{\mathbf{r}} \times \vec{\mathbf{p}} = m\vec{\mathbf{r}} \times \vec{\mathbf{v}} = 0$.

10b. $\vec{\mathbf{L}} = m\vec{\mathbf{r}} \times \vec{\mathbf{v}}, \quad L = mrv\sin\phi = mR_e v$.

11. $F_g = GM_o m/r^2$, where m is the mass of the spacecraft and r is the distance between the Sun and Saturn.

12. $g_T = GM_T/R_T^2$ where M_T and R_T are the mass and radius of Titan respectively.

13. $F_g = GM_J m/r^2$, where r is the distance between the spacecraft and Jupiter's center, M_J is the mass of Jupiter, and m is the mass of the spacecraft.

14a. $p = mv, \quad K = mv^2/2$.
To find the mass of an ammonia molecule NH_3 find its molar mass, i.e. the mass of Avogadro's number of molecules.

14b. $L \approx mR_J v$.

15. $M = \rho V$, where ρ is the density of ice, $V = 4\pi R^3/3$ is the volume of the comet's nucleus.

16. $\rho = M/V$, where M is the mass and V is the volume of the coma.

17a. $S/\Delta t = 4d_a/T$, where d_a is the Sun-comet distance at aphelion, and T is the period of the comet.

17b. $F_a = GM_c M_o/d_a^2, \quad F_p = GM_c M_o/d_p^2, \quad F_a/F_p = (d_p/d_a)^2$, where F_a and F_p are the gravitational forces at aphelion and perihelion, d_a and d_p are the Sun-comet distances at aphelion and perihelion.

18. $M = \rho V$, where M and V are the mass and volume of a fragment.

$$K = Mv^2/2, \quad p = Mv, \quad I = p.$$

19. $F_1 = GmM_J/(r+d)^2$, $\quad F_2 = GmM_J/r^2$, where $r = 20$ mm, $d = 5$ km, and M_J is the mass of Jupiter.

$$F_2 - F_1 \approx 2GmM_J d/r^3 \,.$$

20a. $\rho = m/V$, $\quad V = 4\pi r^3/3$, $\quad r = 5$ mm.

20b. $a = F_g/m = GM_e/(R_e+h)^2 = gR_e^2/(R_e+h)^2$, where $h = 90$ km.

20c. $p = mv$, $\quad K = mv^2/2$.

22. $E = P\Delta t$, $\quad P = 40$ kW/14.

23a. $v = \omega_e(R_e + h)$, where ω_e is the angular speed of the Earth, $\omega_e = 2\pi/T_e$, and $T_e = 1$ day.

23b. $a = v^2/(R_e + h)$.

23c. $F_g = GM_e m/(R_e+h)^2$, $\quad F = ma$, $\quad F = F_g - F_a$, $\quad F_a = F_g - F$, where F_a is the magnitude of the force exerted by the air.

23d. $L = rp$, $\quad L/p = r = (R_e + h)$.

24. $K = mv^2/2$, $\quad p = mv$, $\quad K/p = v/2$.

25. The proton's density is $\rho = m/V$, where m_p is the mass of the proton, $V = 4\pi r_p^3/3$, and r_p is the radius of the proton.

26. $K = (\rho V)v^2/2$, where ρ is the density of air, V its volume ($V = 1$ m^3), and v is the speed of the moving air.

27. $p = (\rho V)v$.

28. $K = mv^2/2$. To find the mass of a nitrogen molecule, N_2, use the Periodic Table of Elements. 1 mol (28 g) of molecular nitrogen contains Avogadro's number of molecules.

29a. $p = mv, \quad K = mv^2/2$.

29b. $I = p$.

30. We consider the motion of Alpha in the system of coordinates connected to the center of the Earth.

$$\frac{GM_em}{r^2} = \frac{mv^2}{r}, \quad r = R_e + h.$$

From this equation
$$v = \sqrt{GM_e/r}, \quad \omega = v/r, \quad T = 2\pi/\omega = 2\pi r/v, \quad a = GM_e/r^2.$$

31. $GM_em/r^2 = mv^2/r, \quad v = \sqrt{GM_e/r}$. If r increases 9 times, v decreases $\sqrt{9} = 3$ times. $\omega = v/r$. If v decreases 3 times and r increases 9 times, then ω decreases 27 times. The acceleration $a = v^2/r$, decreases 81 times. (You can also use $a = GM_e/r^2$.) The period $T = 2\pi/\omega$, increases 27 times.

32. $p = mv$, decreases $\sqrt{9} = 3$ times (see problem 31). $L = rp$, increases 3 times. $K = mv^2/2$ decreases 9 times.

33. $\rho = m/V, \quad m = w/g$, where w is the weight of Destiny.

34a. $F_g = GM_em/r^2, \quad r = R_e + h$.

34b. $p = mv$ (see problem 30).

$$L = rp, \quad K = mv^2/2 = p^2/2m.$$

35. $F_1 = GM_em/r^2, \quad F_2 = GM_om/r_{es}^2, \quad F_1/F_2 = M_e r_{es}^2/M_o r^2$, where $r = R_e + h$, and r_{es} is the mean distance between the Earth and the Sun.

36. $GM_em/r^2 = mv^2/r$, where r is the radius of the geosynchronous orbit. Substitute $r = v/\omega_e$, where $\omega_e = 2\pi/T_e$ is the angular speed of the Earth, $T_e = 1$ day. Thus, $v = (GM_e\omega_e)^{\frac{1}{3}}, \quad p = mv, \quad L = rp, \quad K = mv^2/2$.

37. $T^2/A^3 = 1 (year)^2/(AU)^3$ according to Kepler's Third Law.

38. $V = 4\pi r^3/3$, where r is 10^{10} ly.

39b. $v = \omega r$, $r = 50$ km.

39c. $L = I\omega$, $I = 2Mr^2/5$, $M = \rho V$.

39d. $K = I\omega^2/2$.

41a. $M_c = \rho V$, $V = \pi \cdot 0.4^2 \cdot 1.8$ km^3.

41b. $F_g = GM_eM_c/r^2$.

41c. $\omega = 2\pi/T$.

42. $\dfrac{F_I}{F_D} = \dfrac{M_I}{M_D} = \dfrac{V_I}{V_D}$, where the subscripts "$I$" and "$D$" stand for Ida and Dactyl.

43. $\rho = m_g/V_g$, where m_g and V_g are the mass and the volume of a typical dust grain.

44a. $d = 100$ m.

44b. $V_g/10^6$.

44c. $m_g/10^6$.

44d. $F_g = Gm_g^2/d^2$.

45. $\omega = 2\pi f$, $T = 1/f$, $m = k_s/\omega^2$.

46a. $x = A\cos(\omega t + \phi)$, $v_x = -A\omega \sin(\omega t + \phi)$. At $t = 0$, $x_i = A\cos\phi$, $v_{xi} = -A\omega\sin\phi$. At the end point $v_{xi} = 0$, so $\phi = 0$, or $\phi = \pi$. We choose $\phi = 0$ as $x_i > 0$. Thus, $A = x_i = 40$ nm, and the x-coordinate of the cantilever tip is given by $x = A \cos \omega t$ where A and ω are known. Substituting $t = 1$ ms find the corresponding position of the cantilever tip. For the velocity $v_x = -A\omega \sin \omega t$. For the acceleration $a_x = -A\omega^2 \cos \omega t$ or one can use $a_x = -\omega^2 x$.

46b. $t = T/4, \quad t = 3T/4$.

46c. Consider the equation $x = A \cos \omega t = 20$ nm. It follows that $\cos \omega t = 1/2$. Solutions are $\omega t = \pi/3 + 2\pi k$ and $\omega t = 5\pi/3 + 2\pi k$, $k = 0, 1, 2, \ldots$. The first passage through the point $x = 20$ nm corresponds to the phase $\omega t = \pi/3$, the second passage corresponds to the phase $5\pi/3$. Dividing by ω, find corresponding time.

47a. $v_{\max} = A\omega$.

47b. Use conservation of mechanical energy:

$$\frac{1}{2}k_s x^2 + \frac{1}{2}mv^2 = \frac{1}{2}k_s A^2.$$

Dividing by the mass m one gets

$$(\omega x)^2 + v^2 = (\omega A)^2, \quad v^2 = \omega^2(A^2 - x^2).$$

47c. $a_{\max} = \omega^2 A, \quad F_{\max} = k_s A$.

47d. $a = \omega^2 |x|, \quad F = k_s |x|$.

48. In the equilibrium position the effective spring force balances the magnetic force: $k_s |\Delta x| = F_m$.

49. $v_f = a t_f, \quad t_f = v_f / a$, where v_f and t_f are the final speed and time.

50. $v_f^2 = 2as, \quad s = v_f^2 / 2a$.

51. $F_m = mg$, where F_m is the magnitude of the magnetic force and m is the mass of the train car.

52. $P = F_a v$, $F_a = P/v$, where P is power, and F_a is the force of air resistance.

53. $K = mv^2/2$, $p = mv$.

54. $\frac{dp}{dt} = ma$, dp/dt is the magnitude of the net force acting on the train car.

55a. $v_f^2 = 2as$, $a = v_f^2/2s$.

55b. Let the x-axis point down the incline.

$$F_x = mg\sin\theta - F_e, \quad F_x = ma_x = -ma,$$

$$mg\sin\theta - F_e = -ma, \quad F_e = m(g\sin\theta + a),$$

where F_e is the external force.

56. $s = ((v_i + v_f)/2)t_f = v_f t_f/2$, $t_f = 2s/v_f$.

57a. $v_f = at_f$, where $t_f = 1$ s.

57b. $v_f^2 = 2as$, where $s = 100$ m.

58a. Use the Work–Kinetic Energy Theorem $W_e + W_g + W_r = mv_f^2/2$, where W_e is the work done by the external force, which is equal to energy needed, W_g is the work done by the gravitational force, $W_g = U_i - U_f = -mgs\sin\theta$, and W_r is the work done by the resistive forces (energy dissipation).

58b. $P_{av} = W_e/t_f$, where t_f is the total time of acceleration, which was found in problem 56.

59. Let us start with 59b. At the end of the first second, $v_f = at_f$, where $t_f = 1s$. After the driving force is shut down $v_f' = v_i' - a'(t_f' - t_f)$, where $v_f' = 0$, $v_i' = v_f$, $a' = g\sin\theta$. Thus, $t_f' = t_f + v_f/g\sin\theta$.

59a. The distance covered during the first second, $s = at_f$, where

$t_f = 1s$. The total distance traveled $s' = s + ((v_i' + v_f')/2)(t_f' - t_f)$, where $v_i' = v_f$, $v_f' = 0$.

60a. First consider the motion up the incline. During the first 10 m of acceleration:

$$F_x = -F_e + mg\sin\theta + \mu F_n \, ,$$

where $F_n = mg\cos\theta$, F_e is the external driving force (magnitude), and the x-axis is assumed to point down the incline. Next,

$$F_x = ma_x = -ma \, .$$

Thus,

$$F_e - mg\sin\theta - \mu mg\cos\theta = ma \, ,$$

and we can compute the acceleration a for the first 10 m of travel. The speed of the launcher after the first 10 m, can be found from $v_f^2 = 2as$, where $s = 10$ m.

Now let us consider the motion after the driving force is shut down ($t = 0$) until the moment when the launcher stops. From equation $F_x' = mg\sin\theta + \mu mg\cos\theta = ma'$ we can find the magnitude of the acceleration a', which points down the incline. Next

$$v_f' = v_{x_i}' + a_x't_f, \quad 0 = -v_f + a't_f', \quad t_f' = v_f/a' \, ,$$

$$s' = \frac{v_i' + v_f'}{2}t_f' = \frac{v_i't_f'}{2} = \frac{v_f t_f'}{2} \, ,$$

where t_f' is the time when the launcher stops, s' is the distance covered during that time. For the motion down the incline we have

$$F_x = mg\sin\theta - \mu mg\cos\theta = ma'', \quad (v_f'')^2 = 2a''s'',$$

where $s'' = s' + 10m$. Finally, $s'' = (v_f''/2)(t_f'' - t_f')$. Thus, $t_f'' = t_f' + 2s''/v_f''$.

60b. The distance is $s' + s'' = 2s' + 10$ m.

60c. 10 m

61a. $\Delta t = s/v$, where $s = 4000$ km $= 4 \times 10^6$ m, $v = 150$ m/s.

61b. $T = \lambda/v$.

61c. $f = 1/T$.

62a. $m = \rho A h$ where m is the mass of the water wall, ρ is the density of seawater, h is the height and A is the area of the water wall. $K = mv^2/2$.

62b. $U_g = mgh/2$, $\quad h/2$ is the height of the center of mass of the wall.

62c. $p = mv$.

63. $\Delta t = s/v + (s' + s'')/c$, where v is the speed of sound, c is the speed of light, $s = 2$ km, $s' = 1000$ km, $s'' = 5000$ km.

64a. $\lambda = v/f$.

64b. $T = 1/f$.

64c. $L = v\tau$, where τ is the duration of the pulse, and v is the speed of sound in water.

64d. $N = L/\lambda$, where N is the number of wavelengths in the pulse.

65. x-component of the total angular momentum must be conserved. Thus $L_x = -\hbar/2$.

66. $\vec{p}(\bar{\nu}_\mu) = -\vec{p}(\bar{\nu}_\tau)$.

7.2. Electricity and Magnetism

1. $E = k_c |q|/r^2$, where $q = 2e/3$.

2. $V = k_c q/r$.

3. $F = k_c |q_1 q_2| / r^2$, where $q_1 = 2e/3$, $q_2 = -e/3$.

4. Let the x-axis point in the direction of the electric field. The electric force on the tau lepton has only an x-component, $F_x = qE_x = -eE$, the acceleration of the tau lepton $a_x = -eE/m$. The displacement has only the x-component

$$\Delta x = v_{xi} t + \frac{1}{2} a_x t^2,$$

where $v_{xi} = v_i = 10$ km/s, and the time t is taken to be the mean lifetime of the tau lepton.

5. The same as in problem 4, but $v_{xi} = -v_i$.

6. $C = \epsilon_o A/s$, $V = Q/C$, $E = V/s$, $F = qE = eE$, $a = F/m$. The direction is from the negative plate to the positive plate.

7. $F = qvB$, $\quad q = e$, $\quad a = F/m$.

8. The magnetic force is perpendicular to the velocity. Thus, the muon moves in a circular orbit at constant speed v. The mean distance traveled is vt, where t is the mean lifetime.

9. $U = (k_c/r)(q_1 q_2 + q_1 q_3 + q_2 q_3)$ where q_i are the electric charges of the quarks.

11. $F = k_c e^2 / r^2$, $\quad a = F/m$.

13. $E = k_c q/r^2$, where $q = 2e$.

14. $V = k_c q/r$.

15. $^{209}_{83} Bi$ is the stable isotope with the maximum electric charge. Thus, $F_{\max} = k_c q^2 / r^2$ where $q = 83e$.

16. $\Delta U = 2mB$, where $m = 2.1\mu_n$.

17. $\tau_{\max} = mB$.

18. $B = 2m/r^3$, where $r = 1$ nm, positive z-direction.

19a. $B = 2m/r^3$, positive z-direction.

19b. $B = m/r^3$, negative z-direction.

20. $F = k_c e^2/r^2$.

21. $k_c e^2/r$.

22. $eV = K$, where $V = 1$ MV.

23. Assume that the mass of the fireball is approximately double the mass of a lead nucleus.

$$\rho_1 = m/V_1, \quad \rho_2 = 2m/V_2,$$

$$\rho_2/\rho_1 = 2V_1/V_2 = 2(r_1/r_2)^3 = 20, \quad r_2 = r_1/(10)^{\frac{1}{3}},$$

where m is the mass of the lead nucleus, V_1 is its volume, V_2 is the volume of the fireball. Next,

$$E_1 = k_c q/r_1^2, \quad E_2 = 2k_c q/r_2^2, \quad E_2/E_1 = 2(r_1/r_2)^2,$$

where E_1 is the magnitude of the electric field near the lead nucleus, E_2 is the magnitude of the electric field near the fireball, and q is the electric charge of the lead nucleus.

24. $B = 2m/r^3$, where m is the magnetic moment, $r = 1$ μm.

25. $B_{\max} = 2m_p/(r_o - A)^3$, $\quad B_{\min} = 2m_o/(r_o + A)^3$, $\quad \Delta B = B_{\max} - B_{\min}$, where m_o is the magnetic moment of the magnet, r_o is the equilibrium spacing between the magnet and the atom, and A is the amplitude of the cantilever vibrations.

26a. $m_\mu^{(e)} - m_\mu^{(t)} \approx m_\mu^{(t)} \times 1.3 \times 10^{-6}$ where $m_\mu^{(e)}$ and $m_\mu^{(t)}$ are the experimental and theoretical values of m_μ.

26b. $\Delta U^{(t)} = 2m_\mu^{(t)} B, \quad \Delta U^{(e)} = 2m_\mu^{(e)} B,$

$$\Delta U^{(e)} - \Delta U^{(t)} = 2B(m_\mu^{(e)} - m_\mu^{(t)})$$

27. As shown in Fig. 16 below, the magnetic fields reinforce each other at the bottom and cancel each other at the top.

28. Consider the current $I = 2$ pA in the gap of length 2 nm.

$$\vec{\mathbf{B}}_p \approx \frac{\mu_o}{4\pi} \frac{I \Delta\vec{\mathbf{L}} \times \vec{\mathbf{n}}_p}{r_p^2}$$

where $\Delta\vec{\mathbf{L}} = -2 \times 10^{-9}\vec{\mathbf{k}}$ (in m), $\vec{\mathbf{n}}_p = \vec{\mathbf{i}}x_p/r_p + \vec{\mathbf{j}}y_p/r_p, \quad r_p = (x_p^2 + y_p^2)^{\frac{1}{2}}$.

29a. $\vec{\tau} = \vec{\mathbf{m}} \times \vec{\mathbf{B}}$, where $\vec{\mathbf{m}} = \vec{\mathbf{i}}\mu_B$.

29b. $\Delta U = (\vec{\mathbf{m}}_f - \vec{\mathbf{m}}_i)\vec{\mathbf{B}}$, where $\vec{\mathbf{m}}_i = \vec{\mathbf{i}}\mu_B, \quad \vec{\mathbf{m}}_f = -\vec{\mathbf{i}}\mu_B, \quad \vec{\mathbf{m}}_i$ and $\vec{\mathbf{m}}_f$ are the "initial" and "final" magnetic moments, i.e. the magnetic moments before and after reversal.

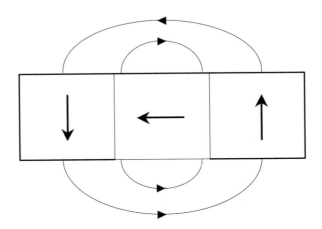

Fig. 16.   Magnetic field lines near the Halbach array.

30. $Q = CV$.

31. $Q = I\Delta t$, $I = jA$, where j is the electric current density, Q is the charge passing through the cross-sectional area A.

32a. $\vec{\mathbf{B}} \approx \dfrac{\mu_o}{4\pi} \dfrac{I\Delta\vec{\mathbf{L}} \times \vec{\mathbf{n}}_p}{r_p^2}$, where $I = jA$ is the current in the wire, $\Delta\vec{\mathbf{L}} = \vec{\mathbf{k}}\Delta L$, ΔL is the length of the segment, the components of $\vec{\mathbf{n}}_p$ are given by $n_{px} = x/r_p$, $n_{py} = y/r_p$, $n_{pz} = z/r_p$, $r_p = (x_p^2 + y_p^2 + z_p^2)^{\frac{1}{2}}$.

32b. $\vec{\mathbf{F}} = q\vec{\mathbf{v}} \times \vec{\mathbf{B}} = -e\vec{\mathbf{v}} \times \vec{\mathbf{B}}$, where $\vec{\mathbf{v}} = \vec{\mathbf{k}}v$.

33a. $m = IA_o = jAA_o$, where $A_o = \pi r_o^2$ is the area of the current loop (do not confuse it with the cross-sectional area of the wire A), r_o is the radius of the loop.

33b. $\vec{\mathbf{B}}_p = \dfrac{\mu_o}{4\pi} \dfrac{3(\vec{\mathbf{m}} \cdot \vec{\mathbf{n}}_p)\vec{\mathbf{n}}_p - \vec{\mathbf{m}}}{r_p^3}$, where $\vec{\mathbf{m}} = \vec{\mathbf{k}}m$ is the magnetic moment of the current loop, the components of $\vec{\mathbf{n}}_p$ are given in the solution to problem 32a.

33c. $\vec{\tau}_a = \vec{\mathbf{m}}_a \times \vec{\mathbf{B}}_p$ where $\vec{\tau}_a$ is the torque on the atom, $\vec{\mathbf{m}}_a = \vec{\mathbf{k}}m_a = 2\mu_B\vec{\mathbf{k}}$ is the atomic magnetic moment.

34a. $|\Delta U| = 2Bm$, where ΔU is the change of the magnetic energy, $B = 1\ T$, $m = IA_o = jAA_o$, A is the cross-sectional area of the wire, and A_o is the area of the square loop.

34b. $\tau = 0$ as the magnetic moment of the current loop is parallel to the magnetic field.

35. The magnetic field (in T) for ion number n is equal to $B_n = 5 + 5 \times 10^{-4}(n-1)$. For the left ion, $n = 1$, for its neighbor $n = 2$, for the right end ion $n = 100$. The energy difference in each case is $\Delta E_n = 2\mu_B B_n$.

36a. $\vec{B}_p = \dfrac{\mu_o}{4\pi} \dfrac{3(\vec{m} \cdot \vec{n}_p)\vec{n}_p - \vec{m}}{r_p^3}$. Let both the center of the magnet and the selected ion lie on the z-axis with the origin at the center of the magnet (see Fig. 13 for the problem). Then

$$\vec{m} = \vec{k}m, \quad \vec{n} = -\vec{k}, \quad \vec{B}_p = \dfrac{\mu_o}{4\pi} \dfrac{2m}{r_p^3}\vec{k},$$

where $\mu_o m / v = 2.2$ T, $V = 4\pi r^3/3$, r is the radius of the magnet, and r_p is the distance between the center of the magnet and the selected ion.

36b. $\tau_{\max} = \mu_B B_p$, where μ_B is the magnetic moment of the ion, and B_p is found in (a).

36c. $\Delta E_{\max} = 2\mu_B B_p$.

37. The magnetic field on the chain has only a z-component (see Fig. 14). Thus, $\vec{B} = \vec{k}B$ for $x > 0$, $\vec{B} = -\vec{k}B$ for $x < 0$, and $\vec{B} = 0$ for $x = 0$. For any point x,

$$B = B_1 - B_2, \quad B_1 = \dfrac{\mu_o I}{2\pi r_1}, \quad B_2 = \dfrac{\mu_o I}{2\pi r_2},$$

where r_1 is the distance between the atom at point x and the center of the nearest wire, r_2 is the similar distance to the second wire, $r_1 = d/2 + \Delta/2 - x$, $r_2 = d/2 + \Delta/2 + x$.

38. $U(0) - U(a) = 3\mu_B B(a)$, $\quad U(a) - U(2a) = 3\mu_B[B(a) - B(2a)]$,

$$U(0) - U(2a) = [U(0) - U(a)] + [U(a) - U(2a)],$$

where $U(0)$, $U(a)$, and $U(2a)$ are the magnetic energies at the positions $x = 0$, $x = a$, $x = 2a$, $B(a)$ and $B(2a)$ are the magnetic fields (magnitude) at $x = a$ and $x = 2a$. Magnetic energy decreases with increase of x, for $x > 0$.

39. $E = P\Delta t$, where P is the power , $\Delta t = 1$ year.

40. $I_r = P/4\pi r^2$, where P is the power radiated by the Sun, r is

the distance from the center of the Sun.

41. $P_r = I_r/c$. Here I_r is the intensity of the Sun's radiation $I_r = P/4\pi r^2$, P is the light power generated by the Sun, r is the distance between the center of the Sun and the asteroid.

42. $F = evB$ where v is the electron speed, B is the magnitude of the magnetic field.

43. $\tau = \mu_B B$, where B is the magnitude of the magnetic field.

44a. $\lambda = c/f$.

44b. $T = 1/f$.

44c. $L = c\tau$, where τ is the duration of the pulse.

44d. $N = L/\lambda$.

45a. $E_i \times (0.7)^n < 0.1 E_i$ where E_i is the initial energy of the pulse, n is the number of reflections. From this equation we get $n > 6.4$, consequently, $n = 7$.

45b. The distance is about $7s$, where s is the separation between the slabs.

45c. $\Delta t \approx 7s/c$.

46. $s = \lambda/2$, where s is the distance between two neighboring traps, λ is the wavelength of the light, $\lambda = c/f$, and f is the frequency of the light.

47a. $L = c\tau$.

47b. $L = c'\tau$, where $c' = 900$ m/s is the speed of the signal light in the gas.

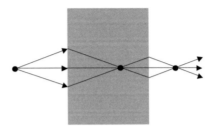

Fig. 17.   Images of the point source inside and outside the slab.

48. $N = s/\lambda$, where N is the number of wavelengths, and s is the distance between the Small Magellanic Cloud and the Sun.

49. The distance to the cluster can be estimated as its size divided by its angular size, which is 10^{10} ly. Thus, the x-rays were emitted 10 billion years ago.

50a. $\lambda = c/f$

50b. $2s$. (See Fig. 17.)

50c. $2w$. (See Fig. 17.)

Chapter 8

CONCLUSION

The purpose of our book is to narrow the gap between the typical introductory course of undergraduate physics and the frontiers of physics. We believe that both undergraduate education and contemporary science would greatly benefit from such a reconciliation. We have chosen one way to approach the goal: to combine in one book selected topics from contemporary science with problems connected to frontier research, which can be solved within the scope of the "standard" undergraduate course. We believe university Professors will use the problems from our book in their courses. We also hope the book will be interesting to undergraduate and graduate students who are curious about contemporary scientific research on a very basic level.

Chapter 9

APPENDICES

9.1. Some Formulas from Vector Analysis

$$\vec{A} \cdot \vec{B} = A_x B_x + A_y B_y + A_z B_z \tag{9.1}$$

$$\vec{A} \cdot \vec{B} = \vec{B} \cdot \vec{A} \tag{9.2}$$

$$\vec{A} \cdot (\vec{B} + \vec{C}) = \vec{A} \cdot \vec{B} + \vec{A} \cdot \vec{C} \tag{9.3}$$

$$\vec{A} \cdot \vec{B} = AB \cos \phi \tag{9.4}$$

$$\vec{i} \times \vec{i} = \vec{j} \times \vec{j} = \vec{k} \times \vec{k} = 0 \tag{9.5}$$

$$\vec{i} \times \vec{j} = \vec{k}, \quad \vec{j} \times \vec{k} = \vec{i}, \quad \vec{k} \times \vec{i} = \vec{j} \tag{9.6}$$

$$\vec{A} \times \vec{B} = -\vec{B} \times \vec{A} \tag{9.7}$$

$$\vec{A} \times (\vec{B} + \vec{C}) = \vec{A} \times \vec{B} + \vec{A} \times \vec{C} \tag{9.8}$$

$$|\vec{A} \times \vec{B}| = AB \sin \phi \tag{9.9}$$

ϕ is the smaller of the angles between the vectors and $|\ldots|$ means the magnitude of the enclosed vector.

9.2. Abbreviations for Units

A	ampere
AU	astronomical unit
C	coulomb
d	day
eV	electronvolt
F	farad
ft	foot, feet
g	gram
h	hour
Hz	hertz
J	joule
K	kelvin
L	liter
lb	pound
ly	lightyear
m	meter
mi	mile
min	minute
mph	miles per hour
N	newton
rad	radian
rev	revolution
s	second
T	tesla
V	volt
W	watt
y	year
yd	yard
°C	degree celsius

9.3. Prefixes

$$\text{exa (E)} = 10^{18}$$
$$\text{peta (P)} = 10^{15}$$
$$\text{tera (T)} = 10^{12}$$
$$\text{giga (G)} = 10^{9}$$
$$\text{mega (M)} = 10^{6}$$
$$\text{kilo (k)} = 10^{3}$$
$$\text{centi (c)} = 10^{-2}$$
$$\text{milli (m)} = 10^{-3}$$
$$\text{micro } (\mu) = 10^{-6}$$
$$\text{nano (n)} = 10^{-9}$$
$$\text{pico (p)} = 10^{-12}$$
$$\text{femto (f)} = 10^{-15}$$
$$\text{atto (a)} = 10^{-18}$$

9.4. Some Conversion Factors

$$1 \text{ AU} = 1.496 \times 10^{11} \text{ m}$$
$$1 \text{ d} = 86400 \text{ s}$$
$$1 \text{ eV} = 1.602 \times 10^{-19} \text{ J}$$
$$1 \text{ ft} = 0.3048 \text{ m}$$
$$1 \text{ h} = 3600 \text{ s}$$
$$1 \text{ L} = 10^{-3} \text{ m}^3$$
$$1 \text{ lb} = 4.448 \text{ N}$$
$$1 \text{ ly} = 9.46 \times 10^{15} \text{ m}$$
$$1 \text{ MeV/c}^2 = 1.78 \times 10^{-30} \text{ kg}$$
$$1 \text{ mi} = 1609 \text{ m}$$
$$1 \text{ y} = 3.156 \times 10^{7} \text{ s}$$
$$1 \text{ yd} = 0.9144 \text{ m}$$

9.5. Approximate Values for Some Physical Quantities

$$\epsilon_o = 8.854 \times 10^{-12} \text{ N/V}^2$$
$$\mu_B = 9.274 \times 10^{-24} \text{ Am}^2$$
$$\mu_n = 5.051 \times 10^{-27} \text{ Am}^2$$
$$\mu_o = 4\pi \times 10^{-7} \text{ N/A}^2$$
$$\rho_{\text{air}} = 1.3 \text{ kg/m}^3$$
$$\rho_{\text{ice}} = 917 \text{ kg/m}^3$$
$$\rho_{\text{seawater}} = 1025 \text{ kg/m}^3$$
$$c = 2.998 \times 10^8 \text{ m/s}$$
$$e = 1.602 \times 10^{-19} \text{ C}$$
$$G = 6.67 \times 10^{-11} \text{ Nm}^2/\text{kg}^2$$
$$g = 9.8 \text{ m/s}^2$$
$$h = 6.626 \times 10^{-34} \text{ Js}$$
$$\hbar = 1.055 \times 10^{-34} \text{ Js}$$
$$k_B = 1.38 \times 10^{-23} \text{ J/K}$$
$$k_c = 8.99 \times 10^9 \text{ V}^2/\text{N}$$
$$m_e = 9.11 \times 10^{-31} \text{ kg}$$
$$m_p = 1.67 \times 10^{-27} \text{ kg}$$
$$N_A = 6.02 \times 10^{23}$$

9.6. Some Astronomical Data

Sun
   Mass: 1.99×10^{30} kg
   Radius: 6.96×10^8 m

Mercury
   Mass: 3.28×10^{23} kg
   Radius: 2.57×10^6 m

Semimajor Axis: 0.387 AU

Period: 88.0 d

Venus

Mass: 4.82×10^{24} kg

Radius: 6.31×10^{6} m

Semimajor Axis: 0.723 AU

Period: 225 d

Earth

Mass: 5.97×10^{24} kg

Radius: 6.38×10^{6} m

Semimajor Axis: 1.00 AU

Period: 1.00 y

Mars

Mass: 6.42×10^{23} kg

Radius: 3.43×10^{6} m

Semimajor Axis: 1.52 AU

Period: 1.88 y

Jupiter

Mass: 1.89×10^{27} kg

Radius: 7.18×10^{7} m

Semimajor Axis: 5.20 AU

Period: 11.9 y

Saturn

Mass: 5.69×10^{26} kg

Radius: 6.03×10^{7} m

Semimajor Axis: 9.54 AU

Period: 29.5 y

Uranus

Mass: 8.66×10^{25} kg

Radius: 2.67×10^{7} m

Semimajor Axis: 19.2 AU

Period: 84.0 y

Neptune
  Mass: 1.03×10^{26} kg
  Radius: 2.48×10^7 m
  Semimajor Axis: 30.1 AU
  Period: 165 y

Pluto
  Mass: 1.1×10^{22} kg
  Radius: 4×10^5 m
  Semimajor Axis: 39.5 AU
  Period: 249 y

Moon
  Mass: 7.35×10^{22} kg
  Radius: 1.74×10^6 m
  Average Earth-Moon distance: 3.84×10^8 m
  Period: 27.3 d

9.7. Notation Used

$\vec{\alpha}$	angular acceleration
β	sound level
β^-	β^--particle (electron)
β^+	β^+-particle (positron)
γ	photon
$\vec{\Delta}\theta$	angular displacement
ϵ_o	permittivity of free space
ϵ	permittivity of an insulating substance
$\vec{\theta}$	angular position
λ	wavelength
μ^-	muon
μ^+	antimuon
μ_o	permeability of free space
μ_B	Bohr magneton

μ_k	coefficient of kinetic friction
μ_n	nuclear magneton
μ_s	coefficient of static friction
ν_μ	muon neutrino
$\bar{\nu}_\mu$	muon antineutrino
ν_τ	tau neutrino
$\bar{\nu}_\tau$	tau antineutrino
ν_e	electron neutrino
$\bar{\nu}_e$	electron antineutrino
π^+	pion
π^-	antipion
$\vec{\tau}$	torque
ρ	density
ρ	resistivity
σ	electric surface charge density
τ^-	tau lepton
τ^+	tau antilepton
ϕ	angle between two vectors
ϕ	phase constant
ϕ_e	electric flux
ϕ_m	magnetic flux
$\vec{\omega}$	angular velocity
ω	angular frequency
A	amplitude
A	area
$\vec{\mathbf{a}}$	acceleration
a_c	magnitude of centripetal component of acceleration
a_t	magnitude of tangential component of acceleration
$\vec{\mathbf{B}}$	magnetic field
b	bottom quark
$\bar{b}$	bottom antiquark
C	capacitance
c	speed of light
c	charm quark
$\bar{c}$	charm antiquark
d	down quark

$\bar{d}$	down antiquark
E	mechanical energy
E_m	magnetic energy
$\mathcal{E}$	electromotive force (EMF)
e	fundamental charge
e^-	electron
e^+	positron(antielectron)
$\vec{\mathbf{E}}$	electric field
$\vec{\mathbf{F}}$	force
$\vec{\mathbf{F}}_g$	gravitational force
$\vec{\mathbf{F}}_n$	normal force
$\vec{\mathbf{F}}_k$	kinetic friction force
$\vec{\mathbf{F}}_s$	static friction force
$\vec{\mathbf{F}}_T$	tension force
f	frequency
G	universal gravitational constant
g	gravitational acceleration at the Earth's surface
h	height above the surface of the Earth
h	Planck's constant
$\hbar$	$h/2\pi$
$\vec{\mathbf{I}}$	impulse
I	electric current
I	intensity
I	moment of inertia
I_o	threshold of hearing
$\vec{\mathbf{i}}$	unit vector in the $+\ x$-direction
$\vec{\mathbf{j}}$	unit vector in the $+\ y$-direction
j	electric current density
$\vec{\mathbf{k}}$	unit vector in the $+\ z$-direction
K	kinetic energy
k	wave number
k_B	Boltzman's constant
k_c	Coulomb's constant
k_s	spring or force constant
$\vec{\mathbf{L}}$	angular momentum

L	length
L_m	inductance
$\vec{\mathbf{m}}$	magnetic moment
M_e	mass of the Earth
M_o	mass of the Sun
m	mass
m_e	electron mass
m_p	proton mass
N_A	Avogadro's number
n	refractive index
n	neutron
$\bar{n}$	antineutron
$\vec{\mathbf{p}}$	momentum
P	power
P_r	radiation pressure
p	proton
$\bar{p}$	antiproton
Q	electric charge on a capacitor
q	electric charge
$\vec{\mathbf{r}}$	position vector
R	resistance
R_e	radius of the Earth
R_o	radius of the Sun
$\vec{\mathbf{S}}$	Poynting vector
S	surface
s	distance traveled by a particle
s	separation of capacitor plates
s	strange quark
$\bar{s}$	strange antiquark
T	period
t	top quark
$\bar{t}$	top antiquark
U	potential energy
U_g	gravitational potential energy
U_s	spring potential energy
u	up quark

$\bar{u}$	up antiquark
u_E	electric energy density
u_B	magnetic energy density
V	volume
V	voltage
V_P	electric potential at the point P
$\vec{\mathbf{v}}$	velocity vector
$\vec{\mathbf{v}}_{av}$	average velocity
W	work

INDEX